Heat and Mass Transfer in Porous Materials

Heat and Mass Transfer in Porous Materials

Editor

Anatoliy Pavlenko

Basel • Beijing • Wuhan • Barcelona • Belgrade • Novi Sad • Cluj • Manchester

Editor
Anatoliy Pavlenko
Department of Building Physics
and Renewable Energy
Kielce University of Technology
Kielce
Poland

Editorial Office
MDPI
St. Alban-Anlage 66
4052 Basel, Switzerland

This is a reprint of articles from the Special Issue published online in the open access journal *Materials* (ISSN 1996-1944) (available at: www.mdpi.com/journal/materials/special_issues/heat_mass_transfer_porous_materials).

For citation purposes, cite each article independently as indicated on the article page online and as indicated below:

Lastname, A.A.; Lastname, B.B. Article Title. *Journal Name* **Year**, *Volume Number*, Page Range.

ISBN 978-3-0365-8763-9 (Hbk)
ISBN 978-3-0365-8762-2 (PDF)
doi.org/10.3390/books978-3-0365-8762-2

© 2023 by the authors. Articles in this book are Open Access and distributed under the Creative Commons Attribution (CC BY) license. The book as a whole is distributed by MDPI under the terms and conditions of the Creative Commons Attribution-NonCommercial-NoDerivs (CC BY-NC-ND) license.

Contents

About the Editor vii

Preface ix

Lukas Maier, Lars Kufferath-Sieberin, Leon Pauly, Manuel Hopp-Hirschler, Götz T. Gresser and Ulrich Nieken
Constitutive Correlations for Mass Transport in Fibrous Media Based on Asymptotic Homogenization
Reprinted from: *Materials* **2023**, *16*, 2014, doi:10.3390/ma16052014 1

Borys Basok, Borys Davydenko, Hanna Koshlak and Volodymyr Novikov
Free Convection and Heat Transfer in Porous Ground Massif during Ground Heat Exchanger Operation
Reprinted from: *Materials* **2022**, *15*, 4843, doi:10.3390/ma15144843 18

Supak Tontisirin, Chantaraporn Phalakornkule, Worawat Sa-ngawong and Supachai Sirisawat
Magnetic Induction Assisted Heating Technique in Hydrothermal Zeolite Synthesis
Reprinted from: *Materials* **2022**, *15*, 689, doi:10.3390/ma15020689 32

Krzysztof Blauciak, Pawel Szymanski and Dariusz Mikielewicz
The Influence of Loop Heat Pipe Evaporator Porous Structure Parameters and Charge on Its Effectiveness for Ethanol and Water as Working Fluids
Reprinted from: *Materials* **2021**, *14*, 7029, doi:10.3390/ma14227029 44

Joanna Grzelak and Ryszard Szwaba
Influence of Holes Manufacture Technology on Perforated Plate Aerodynamics
Reprinted from: *Materials* **2021**, *14*, 6624, doi:10.3390/ma14216624 61

Konstantin Kappe, Michael Bihler, Katharina Morawietz, Philipp P. C. Hügenell, Aron Pfaff and Klaus Hoschke
Design Concepts and Performance Characterization of Heat Pipe Wick Structures by LPBF Additive Manufacturing
Reprinted from: *Materials* **2022**, *15*, 8930, doi:10.3390/ma15248930 72

Katarzyna Skrzypczyńska, Andrzej Świątkowski, Ryszard Diduszko and Lidia Dąbek
Studies on Carbon Materials Produced from Salts with Anions Containing Carbon Atoms for Carbon Paste Electrode
Reprinted from: *Materials* **2021**, *14*, 2480, doi:10.3390/ma14102480 90

Alexander Shkarovskiy and Shirali Mamedov
Improving the Efficiency of Non-Stationary Climate Control in Buildings with a Non-Constant Stay of People by Using Porous Materials
Reprinted from: *Materials* **2021**, *14*, 2307, doi:10.3390/ma14092307 100

Borys Basok, Borys Davydenko and Anatoliy M. Pavlenko
Numerical Network Modeling of Heat and Moisture Transfer through Capillary-Porous Building Materials
Reprinted from: *Materials* **2021**, *14*, 1819, doi:10.3390/ma14081819 109

Krzysztof Kuśmierek, Andrzej Świątkowski, Katarzyna Skrzypczyńska and Lidia Dąbek
Adsorptive and Electrochemical Properties of Carbon Nanotubes, Activated Carbon, and Graphene Oxide with Relatively Similar Specific Surface Area
Reprinted from: *Materials* **2021**, *14*, 496, doi:10.3390/ma14030496 130

About the Editor

Anatoliy Pavlenko

Professor Anatoliy Michael Pavlenko is Head of the Department of Building Physics and Renewable Energy of Kielce University of Technology, Poland. Scientific direction of work: thermophysics of dispersed media. Scientific interests: mathematical modeling of the thermophysical processes occurring in liquids in the metastable state; using the CFD approach to combustion modelling and numerical simulation of thermal processes; use of the finite element technique in simultaneous transient conduction; and thermal stress analysis.

Preface

Taking into account that almost all technologies are associated with porous media and heat and mass transfer processes in them, both in technological processes and in nature, the field of the scientific research of PM is practically unlimited. Porous media play an important role in a wide range of scientific and engineering problems. Therefore, the problems of their application are associated with the solution of multiscale processes that combine the transfer of mass, momentum, and energy.

Heat and mass transfer processes in PM are part of our daily experience, and this overall process is central to many environmental and engineering applications, from the evaporation of moisture from the soil to the drying of various products and building materials.

Thus, heat and mass transfer in porous media has been an important research topic for many decades because of its applicability. Despite numerous studies spanning over a century, many new discoveries are still emerging that improve the basic understanding of the subject.

Anatoliy Pavlenko
Editor

Article

Constitutive Correlations for Mass Transport in Fibrous Media Based on Asymptotic Homogenization

Lukas Maier [1,*], Lars Kufferath-Sieberin [1], Leon Pauly [2], Manuel Hopp-Hirschler [1], Götz T. Gresser [2,3] and Ulrich Nieken [1]

1. Institute of Chemical Process Engineering, University of Stuttgart, Boeblinger Strasse 78, 70199 Stuttgart, Germany
2. German Institutes of Textile and Fiber Research Denkendorf (DITF), Körschtalstraße 26, 73770 Denkendorf, Germany
3. Institute for Textile and Fiber Technologies (ITFT), University of Stuttgart, Pfaffenwaldring 9, 70569 Stuttgart, Germany
* Correspondence: lukas.maier@icvt.uni-stuttgart.de

Abstract: Mass transport in textiles is crucial. Knowledge of effective mass transport properties of textiles can be used to improve processes and applications where textiles are used. Mass transfer in knitted and woven fabrics strongly depends on the yarn used. In particular, the permeability and effective diffusion coefficient of yarns are of interest. Correlations are often used to estimate the mass transfer properties of yarns. These correlations commonly assume an ordered distribution, but here we demonstrate that an ordered distribution leads to an overestimation of mass transfer properties. We therefore address the impact of random ordering on the effective diffusivity and permeability of yarns and show that it is important to account for the random arrangement of fibers in order to predict mass transfer. To do this, Representative Volume Elements are randomly generated to represent the structure of yarns made from continuous filaments of synthetic materials. Furthermore, parallel, randomly arranged fibers with a circular cross-section are assumed. By solving the so-called cell problems on the Representative Volume Elements, transport coefficients can be calculated for given porosities. These transport coefficients, which are based on a digital reconstruction of the yarn and asymptotic homogenization, are then used to derive an improved correlation for the effective diffusivity and permeability as a function of porosity and fiber diameter. At porosities below 0.7, the predicted transport is significantly lower under the assumption of random ordering. The approach is not limited to circular fibers and may be extended to arbitrary fiber geometries.

Keywords: homogenization; textiles; yarns; random fibers; permeability; effective diffusion

1. Introduction

Textiles are omnipresent in industrial and everyday applications, such as clothing, composite materials and construction materials. Textiles are an important class of technical materials due to their great flexibility in shape and inexpensive production processes. The performance and area of application of a particular textile often depend on its permeability. Current examples of the importance of mass transport in textiles are sportswear [1] and woven gas diffusion layers in fuel cells [2]. In sportswear, the wearing comfort is strongly influenced by air permeability and water vapor diffusion through the textile [3]; in fuels cells, the water transport in the gas diffusion layer influences the fuel cell performance [4]. Continuous filament synthetic yarns such as polyester and carbon are often used in the above applications. To adapt and optimize a given textile according to the required specifications of its application, a basic knowledge of the transport within the textile is necessary. Today, the mass transport is usually determined in experiments. However, it is not possible to distinguish between the influence of the yarn and the structure of the knitted or woven fabric on mass transfer. To improve the mass transport properties of yarns, it is desirable to study the individual influence of yarn and fabric structure separately.

Therefore, correlations based on the assumption of a regular arrangement of fibers (hexagonal and square lattice) are often used to predict the permeability of yarns [5–7]. However, this leads to an overestimation of the mass transfer properties [8]. In this contribution, we present new correlations for estimating the effective diffusivity and permeability of yarns as a function of porosity and fiber diameter, which take random ordering into account. To this end, we consider yarns made from continuous filaments of synthetic material. Here, the term porous or fibrous media will only refer to the yarn.

The convective and diffusive mass transport in the fibrous (porous) media can be predicted in a pore-scale simulation where the transport equations are solved on a representative cutout from μCT-images or FIB-Sem images [9]. However, pore-scale simulations are computationally very expensive and, hence are not suitable to simulate large areas of yarn material. Therefore, porous media are considered to be an effective media and transport is lumped into effective transport properties. In the context of textiles, the yarn can also be treated as an effective medium and only the weave or knit geometry is resolved in detail to simulate the mass transport in textile materials [5,10].

To model the effective convective transport in a fibrous material and in general porous media, the well-known Darcy equation is commonly applied [11]:

$$v = -\frac{K}{\eta} \cdot \nabla p, \qquad (1)$$

where v, K, ∇p and η are the volume-averaged flow velocity, permeability tensor, pressure gradient and the dynamic viscosity of the fluid, respectively. The permeability K reflects the microstructure of the porous media, since all the convective microscale transport phenomena are mapped in the permeability K. Therefore, it is of high interest to determine the permeability K based on the structural information of the fibrous (porous) media. This relationship is valid in the creeping-flow regime (Reynolds Number $\ll 1$).

One of the key structural parameters determining the permeability of fibrous materials—indeed, all porous materials—is the porosity $\phi = V_{pore}/V_{media}$, where V_{media} is the total volume of the fibrous media and V_{pore} is the volume not occupied by the fibers [12]. Obtained by gravimetrical measurement or imaging analysis, it is also the easiest structural parameter to identify [13]. Thus, it is of great interest to determine a constitutive permeability–porosity correlation of yarns or tows. Several researchers have published relationships of the permeability of fibrous materials as a function of their porosity for ordered fibrous media in two-dimensions or randomly oriented fiber networks in three-dimensions [8,14–20].

Gebart et al. [8] present an analytical, experimental and numerical investigation of the permeability of hexagonal and squared-lattice-ordered two-dimensional arrays of fibers. They derive the following correlation for $K(\phi)$:

$$\frac{K}{r^2} = C \left(\sqrt{\frac{1 - \phi_{per}}{1 - \phi}} - 1 \right)^{\frac{5}{2}}, \qquad (2)$$

where r is the fiber radius, ϕ_{per} is the critical value of porosity below which there is no permeating flow (the percolation threshold) and C is a geometric factor $C = 16/\left(9\pi\sqrt{2}\right)$ and $\phi_{per} = 1 - \pi/4$ for a squared arrangement, and $C = 16/\left(9\pi\sqrt{6}\right)$ and $\phi_{per} = 1 - \pi/2\sqrt{3}$ for the hexagonal-arranged fibers. The obtained correlation shows an excellent agreement to the numerical results. Since the analytical consideration assumes that the permeability is controlled by the narrow slots between the fibers, the correlation is only valid for a maximal porosity $\phi = 0.65$, according to Gebart et al. [8].

To study creeping flow through three-dimensional random fiber packings such as non-woven fabrics or paper-like materials, Koponen et al. [19] used the lattice Boltzmann method (LBM). Clague et al. [18] and Nabovati et al. [14] also studied the permeability of three-dimensional ordered and disordered fibrous media. They used the LBM to simulate creeping flow through fully three-dimensional random fiber networks, in which overlapping of the fibers was allowed. Based on the LBM simulations, a permeability correlation was proposed. Nabovati et al. fitted the numerical results to the constitutive permeability correlation proposed by Gebart et al. [14]. Based on asymptotic homogenization, Schulz et al. [20] studied the permeability for squared-lattice-ordered two-dimensional porous media with different geometries, such as ellipses, squares and rectangles. With these results they proposed new permeability correlations that expand the famous Kozeny–Carman equation to different geometries, since the original correlation assumes spherical particles. However, to the best knowledge of the authors, there is no constitutive permeability correlation for yarns or tows, which considers random parallel-arranged fibers.

To model the diffusive transport of a species in a fibrous media, the Ficks–Diffusion law is widely applied [21]

$$J = D_{\text{eff}} D \nabla \bar{c}, \qquad (3)$$

where J, D_{eff}, D and $\nabla \bar{c}$ are the diffusive flux, effective diffusion coefficient, bulk diffusion coefficient and volume averaged concentration gradient, respectively. The bulk diffusion coefficient might be concentration-dependent, or dependent on the pore size in the case of Knudsen diffusion [22,23]. The effective diffusion coefficient is a structural parameter that considers the transport hinderance induced by the pore structure. For yarns and tows, it is of interest to determine a constitutive effective diffusivity–porosity correlation. Several previously published articles describe the relationships of the effective diffusion coefficient of fibrous materials as a function of their porosity for ordered fibrous media in two dimensions [24–31].

Perrins et al. [28] derived an analytical correlation for $D_{\text{eff}}(\phi)$ for hexagonal and squared-lattice-ordered two-dimensional arrays of fibers, by applying the method of Lord Rayleigh [32]:

$$D_{\text{eff}} = \frac{1}{\phi}\left(1 - \frac{2(1-\phi)}{2 - \phi - C_1(1-\phi)^{C_2}}\right) \qquad (4)$$

where ϕ is the porosity and C_1 and C_2 are geometric factors, where $C_1 = 0.3058$ and $C_2 = 4$ for a square array, and $C_1 = 0.07542$ and $C_2 = 6$ for a hexagonal lattice. The correlation was validated numerically with different methods, such Monte Carlo simulations, asymptotic homogenization, the Voronoi tessellation method with mixing rules and by several researchers [24,25,29].

In actual yarns or tows, the fibers are not arranged in a square or hexagonal lattice, which is a basic assumption for correlations (2) and (4). In this contribution, we want to adapt the presented correlations to a randomly placed fiber setting. To do so, we apply a mathematical upscaling method.

Mathematical upscaling methods can be used to compute effective transport coefficients [33]. Helmig et al. and Battiato et al. provide a comprehensive overview of upscaling techniques [34,35]. These techniques derive macroscopic transport equations from first principles. The methods have been developed by different academic communities, such as mathematicians and engineers, and by using different methodologies. However, all methods lead to the same result for standard transport phenomena, such as diffusion and creeping single-phase flow in porous media considered here.

Asymptotic homogenization is a well-known upscaling technique based on asymptotic expansions and can be used to determine effective transport parameters for porous media. The local transport processes occurring within the porous material, as well as structural data such as porosity, influence the effective transport coefficients derived by this method. Here, we compute by asymptotic homogenization effective transport coefficients for a set of randomly generated unit cells. Curve fitting of our numerical results revealed a constitutive relationship for the permeability and the effective diffusion coefficient as a function of porosity. A similar approach has been used by Kamiński et al. and Jeulin et al. in the context of continuum mechanics and acoustics to estimate the effective material properties of composites [36,37].

The paper is structured as follows. Section 2 introduces the methodology of asymptotic homogenization and summarizes the convective and diffusive mass transport equations at the pore and continuum scale based on the homogenization theory. Additionally, the methodology of deriving a constitutive permeability–porosity correlation is presented. In Section 3, the numerical results and constitutive correlations are presented. Section 4 summarizes the results and gives an outlook on work in progress.

2. Methodology

The asymptotic homogenization is a mathematical averaging method that can be used to derive macroscopic transport equations, stating from a microscale description [38–40]. To use this method, a periodic representation on a microscopic scale must exist to represent the heterogenous media [41]. As explained in the next section in more detail, the prerequisite of periodicity can be relaxed in practical application. In this contribution, the periodic representation is denoted as Representative Volume Element Y. A cornerstone of the homogenization is the scale separation between the macroscale and the microscale. This is expressed by the size difference between the microscopic scale l_c and the macroscopic scale L_c:

$$\varepsilon = \frac{l_c}{L_c} \ll 1. \tag{5}$$

when ε is small, the asymptotic expansion with respect to ε can be applied to the microscale description of the transport phenomena. The asymptotic expansion is

$$a(x)^\varepsilon = a_0(x,y) + \varepsilon a_1(x,y) + \varepsilon^2 a_2(x,y) + \cdots. \tag{6}$$

$a(x)^\varepsilon$ is spatially varying on the microscale y. $a(x)^\varepsilon$ is substituted into the equation describing the transport phenomenon on the microscale and can represent, e.g., a concentration or a velocity. The lower index here indicates the hierarchy of the scale; a_0 refers to the macroscale, a_1 to the next smaller scale, and so on. Equation (6) states that the scale becomes smaller and smaller as the index rises. In two scale asymptotic homogenization, order terms higher than ε^2 are neglected.

The limit of homogeneity is reached when ε goes to zero, at which point the heterogeneity becomes infinitely fine, as shown in Figure 1. Since there are no more structural changes in the microscopic variable and the domain is homogeneous at $\varepsilon \to 0$, the equation no longer depends on the microscopic variable y. By determining the limit $\varepsilon \to 0$, the effective transport equation of a physical process in heterogeneous media is derived by asymptotic analysis.

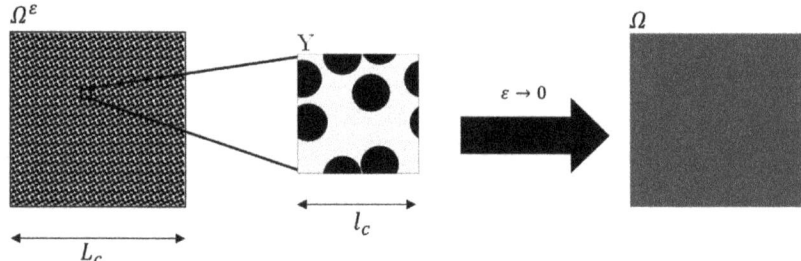

Figure 1. Conceptualization of the asymptotic homogenization.

In the following section, we apply the asymptotic homogenization to convection and diffusion in fibrous media.

2.1. Representative Volume Element

In addition to a clear scale separation, the second prerequisite for the application of asymptotic homogenization is the existence of a spatially periodic domain that is representative of the porous or fibrous media. In the context of volume averaging—another upscaling method—this representative domain is also referred to as a Representative Volume Element (RVE). For more information on the concept of RVE, we refer to the classical literature [11,12,42]. As shown by several researchers, the periodicity is not a strict requirement in the sense that the structure must be strictly periodic. They have shown that a slow variation of the structural parameters over the macroscopic length still allows the application of asymptotic homogenization to derive transport parameters for porous media [22,23,43]. It has been shown also that it is possible to represent the real porous structure, and thus non-strictly spatially periodic material by artificially generated RVEs, which solely represented the characteristic structural parameters such as porosity, pore-size and width of the pore throat of the real porous media, without representing the porous structure in detail [22,43–45].

Based on this concept, we propose a schematic representation as shown in Figure 2. The RVE is a square domain that contains randomly periodically arranged circles. In three dimensions this can be considered as parallel fibers. By choosing a fiber diameter and fiber number, the porosity can be set. We limit the investigation to uniform nonoverlapping circular fibers, as they are most commonly found in technical textiles or composites manufactured from endless filaments from synthetic materials. Moreover, we only consider the transport perpendicular to the fibers, since for sportswear and clothing the air permeability and the water vapor transmission is usually determined perpendicular to the fibers. Our geometrical simplification is underpinned by microscopic visualization and research works of modelling water transport in yarns [13,46,47]. For yarns with twist, a three-dimensional RVE may be necessary since the transport along the fibers is influenced by the twist angle. However, we want to emphasize that the approach can be easily extended to different geometries by choosing different fiber cross-sections such as ellipses, squares or fiber-size distributions, and even a three-dimensional RVE is possible.

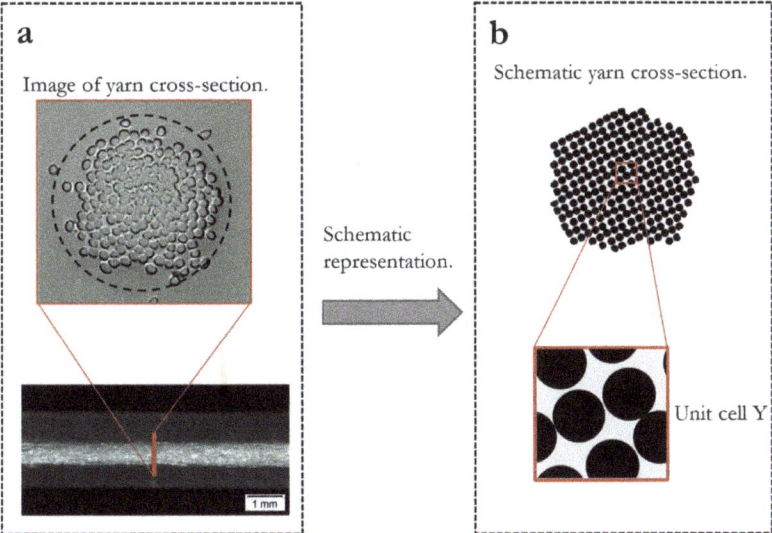

Figure 2. (**a**) Image of a melt spun multifiber polyester yarn and visualization of a cross-section of yarn by embedding in resin. (**b**) Schematic representation of yarn cross-section with corresponding RVE Y.

The RVE Y is generated using Matlab®, according to the workflow presented in Figure 3.

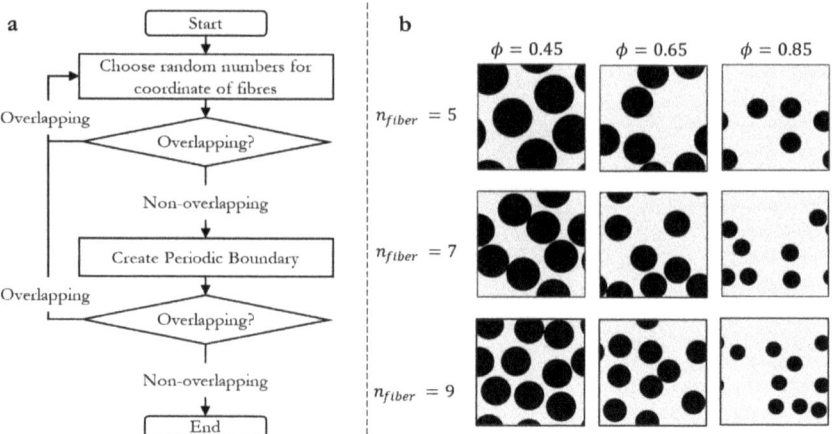

Figure 3. (**a**) Schematic process diagram of the generation of RVEs. (**b**) Exemplary RVEs with different porosities and fiber numbers.

First, $2 \cdot n_{fibers}$ random numbers are chosen in a range from 0 to $l_{RVE} + 2r$, where n_{fibers} is the number of fibers, l_{RVE} is the sidelength of the RVE and r is the fiber radius. The random numbers are generated using the Mersenne–Twister algorithm and are uniformly distributed between 0 to $l_{RVE} + 2r$ [48]. We have chosen $l_{RVE} = 1$. The radius of the fibers is calculated to match the desired porosity ϕ for a given number of fibers. Each number duple is a Cartesian coordinate of the center of a fiber. If the fibers did not overlap, periodic boundaries were created, in order to apply periodic boundaries in the simulation. This was performed by checking whether the fibers violated the boundaries of the domain. If the vertical boundaries were crossed by a fiber, the fiber was mirrored by adding (for the left boundary) or subtracting (for the right boundary) l_{RVE} to the x-value of the Cartesian coordinate. The same algorithm was applied to the horizontal boundaries. The Euclidean distance between the fibers was calculated subsequently. If an overlap between the fibers was observed, the procedure was started again. Due to the relatively small number of fibers in the RVE, the creation of the RVE is not very time-consuming. With this rather simple algorithm, a statistically random arrangement of the fibers in the RVE can be realized. The RVE was generated using COMSOL Multiphysics®'s built-in CAD functionality and meshed in COMSOL Multiphysics® based on the fiber-position data generated in Matlab®. In Figure 3, examples of RVEs with different numbers of fibers and porosities are shown.

In the next sections, we summarize the formal asymptotic homogenization of convective and diffusive transport in porous media.

2.2. Homogenization of Convective Transport

The convective mass transport on the pore-scale is covered by the Navier–Stokes equations. Due to the small pores of the fibrous media and the low velocity, creeping flow (Reynolds Number $\ll 1$) is a reasonable assumption, i.e., inertia can be neglected.

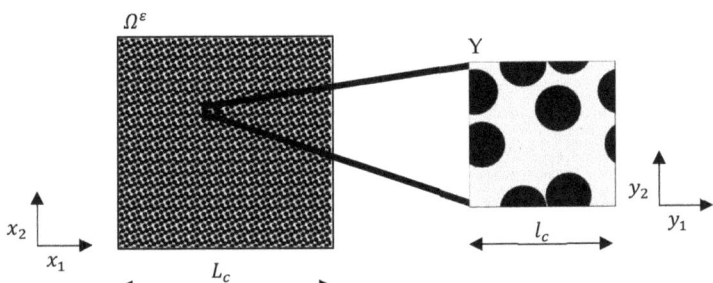

Figure 4. Fibrous media that is composed of spatially periodic RVE. The RVE consists of randomly arranged circles representing the fibers in two dimensions.

For a fibrous (porous) medium with impermeable walls (see Figure 4), this gives the Stokes problem the following form:

$$\begin{cases} \varepsilon^2 \eta \Delta_y v^\varepsilon - \nabla_y p^\varepsilon = 0, & \text{in } Y_F \\ \nabla_y v^\varepsilon = 0, & \text{in } Y_F \\ v^\varepsilon = 0, & \text{on } \Gamma. \end{cases} \quad (7)$$

where v^ε, p^ε, η, Y_F, ε are the velocity, pressure, viscosity, pore space and the order parameter of scale separation, respectively. Slip on the wall Γ of the fibers and additional body forces are neglected. In order to perform an asymptotic expansion, the viscosity is scaled by ε^2. Without scaling, frictional forces would dominate for $\varepsilon \to 0$. This means the pressure gradient would have no effect on the velocity profile. Physically, this is reflected by standstill [49]. The scaling postulates that the shear forces are in equilibrium with the frictional forces. This results in a physically reasonable solution to the problem for $\varepsilon \to 0$.

In the next step, an asymptotic expansion for the quantities v^ε and p^ε is performed. For this, the ansatz (6) and (7) are applied to the pore-scale transport Equation (8). This leads to the following cell problems in Y for convective mass transfer:

$$\begin{cases} \Delta_y \vec{\chi}_j - \nabla_y \Pi_j + \vec{e}_j = 0, & y \in Y_F \\ \nabla_y \vec{\chi}_j = 0, & y \in \Gamma \\ \vec{\chi}_j = 0, & y \in \Gamma \\ \Pi_j \text{ and } \vec{\chi}_j \text{ are } Y - periodic, \end{cases} \quad (8)$$

where the base functions $\vec{\chi}_j$ and Π_j are the local variation of the velocity and pressure, the lower index j denotes the spatial directions, e.g., in two-dimensions $j = [1, 2]$. Thus, two cell problems are solved to determine the permeability tensor in two dimensions. In addition, we assume that the porous media Ω^ε is composed of a spatially periodic RVE Y. The detailed derivation of the cell problem (9) can be found in [49].

The result of the two-scale asymptotic expansion is the Darcy equation:

$$v = -\frac{K}{\eta} \cdot \nabla p, \quad (9)$$

where v, K, ∇p and η are the volume-averaged flow velocity, permeability tensor, pressure gradient and the dynamic viscosity of the fluid, respectively. The permeability can be calculated by solving the cell problems (8) on Y (see Figure 3). By volumetric averaging of the base functions $\vec{\chi}_j$:

$$k_{ij} = \frac{1}{|Y|} \int_{Y_F} \chi_{ij} dy, \quad (10)$$

where Y_F is the void space, the permeability tensor in, e.g., two dimensions is given by:

$$K = \begin{pmatrix} k_{11} & k_{12} \\ k_{21} & k_{22} \end{pmatrix}. \tag{11}$$

To transform the permeability K in a dimensionless form, K is scaled by the square of the characteristic microscopic length K/l_c^2. We choose as the microscopic characteristic length $l_c = r$, where r is the fiber radius. To determine the permeability tensor K, the cell problem (8) in the RVE Y must be solved first.

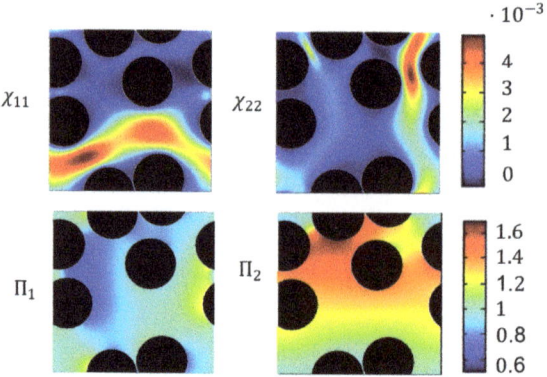

Figure 5. Base functions of the cell problem for convection in the RVE geometry with five fibers. On the left side, the corresponding base functions to the y_1-direction, and on the right side to the y_2-direction. **Top**: Base functions velocity χ_{ij}. **Bottom**: Base functions pressure Π_j.

An example of the base functions is shown in Figure 5. The simulation of the cell problem was performed with the finite-element simulation software COMSOL Multiphysics®. On the boundaries of the domain, we applied periodic conditions. For discretization, a triangular mesh was applied, where the $\vec{\chi_j}$ variable was discretized with second order elements, while Π_j was discretized with first order elements. The direct solver PARADISO was used to solve the cell problem [50]. The permeability tensor was calculated by surface integration. Numerical accuracy was ensured by a mesh study not shown here.

2.3. Homogenization of Diffusive Transport

In this section, we outline the homogenization of diffusive transport. Diffusion in the pores Ω_F^ε of the periodic porous media Ω^ε, illustrated in Figure 4, can be modelled by the pore-scale transport equation:

$$-\nabla \cdot (D \nabla c(x)) = 0 \ in \Omega_F^\varepsilon. \tag{12}$$

D is the molecular diffusion coefficient and $c(x)$ is the concentration that is depended on the macroscopic spatial variable x. To use the asymptotic homogenization to derive an effective diffusion equation, we scale the variables by characteristic quantities: $D^* = D/D_c$, $c^* = c/c_c$, $y^* = y/l_c$ and $x^* = x/L_c$. The parameters with subscript c are the characteristic parameters, respectively. Here, we choose the molecular diffusion coefficient D as characteristic diffusion coefficient D_c. Dropping the asterisks, the scaling results in the dimensionless equations are as follows:

$$\begin{cases} -\nabla \cdot (D^\varepsilon \nabla c^\varepsilon) = 0, & in \ \Omega_F^\varepsilon \\ -\vec{n} \cdot (D^\varepsilon \nabla c^\varepsilon) = 0, & on \ \Gamma^\varepsilon \\ c^\varepsilon = c_D, & on \ \partial\Omega^\varepsilon, \end{cases} \tag{13}$$

where $\vec{n}$ is the normal vector on the pore wall Γ^ε and D^ε is the dimensionless diffusion tensor. The index ε indicates the dependence on the microscopic variable y.

By applying the expansion (6) to the pore-scale transport Equation (13), the base functions $w_j(y)$ are the self-similar local changes in concentration and the so-called cell problem. For a more detailed derivation, we refer to [49,51–53].

$$\begin{cases} -\Delta_y w_j = 0, & y \in Y_F \\ \vec{n} \cdot (\nabla_y w_j) = -\vec{n} \cdot \vec{e_j}, & y \in \Gamma \\ w_j \text{ is } Y - \text{periodic}. \end{cases} \quad (14)$$

The result of the homogenization after redimensioning is an effective diffusion equation

$$-\nabla \cdot \left(D_{eff} D \nabla \bar{c} \right) = 0 \text{ in } \Omega \quad (15)$$

D_{eff} is the effective dimensionless diffusion tensor, with

$$d_{ki} = \frac{1}{|Y|} \int_{Y_F} \left(\delta_{ki} + \frac{\partial}{\partial y_k} w_j(y) \right) dy, \quad (16)$$

where w_j is the solution to the cell-problem (14). Since w_j is integrated via Y_F, the integral refers to the averaged concentration in the void space c_i. To relate the diffusion coefficients to the volume-averaged concentration, the coefficient must be scaled by $\frac{1}{\phi}$, since $c_i = \phi \cdot \bar{c}_i$. The effective diffusion tensor then reads:

$$D_{eff} = \frac{1}{\phi} \cdot \begin{pmatrix} d_{11} & d_{12} \\ d_{21} & d_{22} \end{pmatrix} \quad (17)$$

with the averaged concentration $\bar{c}_i$ of the species i in the porous media Ω.

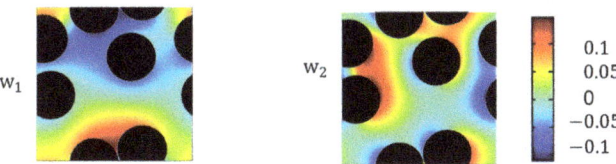

Figure 6. Base functions of diffusion in the RVE geometry with five fibers. w_1 is the base function of the y_1-direction and w_2 to the y_2-direction.

The computed basis functions w_j are shown in Figure 6. As for the cell problem of convective transport, we implemented the cell problem (15) in the commercial simulation program COMSOL Multiphysics® to solve the set of equations, and the same solver and boundary conditions were applied. For discretization, a triangular mesh was applied, where the w_j base functions were discretized with second order elements. By numerical surface integration, the diffusion tensor was derived.

2.4. Deriving Constitutive Transport Relationships Based on Asymptotic Homogenization

In the following, we outline the approach to derive constitutive transport relationships. A schematic flow chart of the methodology is shown in Figure 7.

In the first step, the virtual RVEs are created in order to represent the characteristic microstructure of the considered material. For a fixed porosity, a representative number of RVEs are created to represent the average geometrical configuration. In the next step, the cell problems (8) and (15) are solved on a set of RVEs, which are randomly generated taking porosity and fiber diameter as a constraint. Finally, the transport properties obtained by the simulations are fitted to correlations (2) and (4). This averages the transport coefficients over all random geometries at a given porosity ϕ.

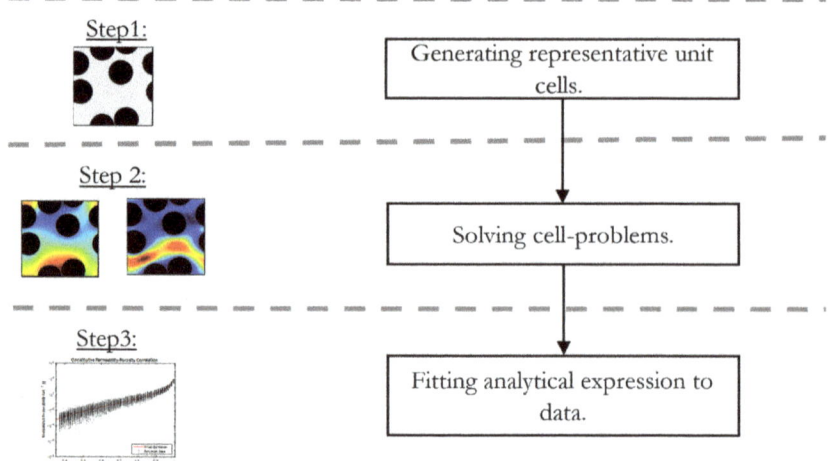

Figure 7. Schematic flow chart to determine constitutive correlations for mass transport in fibrous media.

3. Results and Discussion

In this section, we present the simulation results and improved constitutive correlations for mass transport in fibrous media. To demonstrate the impact of random ordering, we compare our proposed correlations to correlations by Gebart [8].

3.1. Number of Fibers in the Representative Volume Element

To determine a representative and generally valid correlation, it is of great importance that the numerical results must be invariant to the number of fibers in the RVE. To verify this, 300-RVEs were generated for different porosities $\phi = 0.45$ to $\phi = 0.95$ with different number of fibers ($n_{fiber} = [5, 6, 7, 8]$), giving in total 1500-RVE. Next, the data were averaged for the respective porosity at a fiber number. As pointed out by King et al. [54], transport coefficients for random porous media have been geometrically averaged. As can be seen in Figure 8, even five fibers are sufficient to obtain numerical results, which are independent of the number of fibers and therefore the size of the RVE.

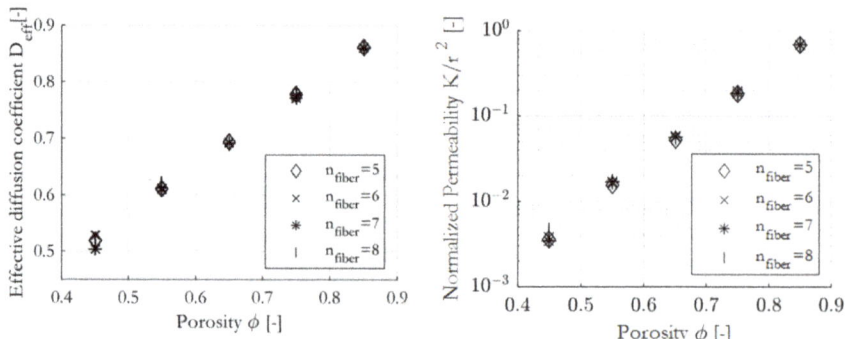

Figure 8. Geometrical mean of transport coefficients, computed for RVEs with a different number of fibers.

In all calculations, we set the number of fibers in the RVE to be $n_{fiber} = 5$ to minimize the computational cost.

3.2. Constitutive Permeability–Porosity Relationship for Random Arranged Parallel Fibers

We determine the permeability of 300 random RVEs in the x- and y-direction for each given porosity. In all RVEs, the fibers have the same radius. The porosity of the RVE varied from 0.35 to 0.99. Since the geometry is isotropic on average, 600 data points were obtained for each porosity; hence, we received 38,400 data points in total. The results are plotted in Figure 9. As expected, the permeability tends to infinity in the limit $\phi \to 1$ and drops towards zero at low porosity. The variation in the data is greatest at low porosities, where the random arrangement of the fibers leads to large relative changes in the predicted permeability. A modified version of the Gebart [8] relationship provided an excellent fit over the range of porosities considered. We adapt Gebart's original relationship by adapting three constants:

$$\frac{K}{r^2} = C_1 \left(\sqrt{\frac{1 - \phi_{per}}{1 - \phi}} - 1 \right)^{C_2}, \quad (18)$$

where ϕ_{per} is the value of porosity above which flow can occur, in fact the percolation threshold. C_1 and C_2 relate to the RVE geometry. A similar approach was chosen by Nabovati et al. [14]. The fitted correlation is shown in Figure 9.

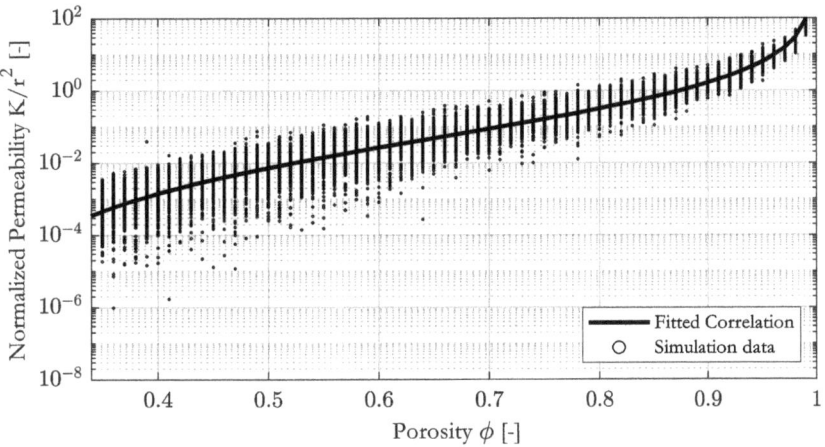

Figure 9. Constitutive permeability–porosity correlation and simulation data.

We used the lsqcurvefit function of Matlab®, with a Levenberg–Marquardt algorithm for minimizing the Euclidean norm. The fit is performed in logarithmic space to avoid biasing the fit towards large permeability values at high porosity. The parameters are given in Table 1.

Table 1. Fit values of the permeability–porosity correlation.

Parameter	Bestfit Value
C_1	0.3468
C_2	2.6193
ϕ_{per}	0.2306

The fitted ϕ_{per} is comparable in magnitude to the analytically determined percolation threshold of the squared arrangement ($\phi_{per} = 0.21$) by Gebart [8], but due to the random arrangement in our geometrical consideration, the percolation threshold reached a higher porosity, as expected. However, we emphasize our correlation is only valid in a porosity range from 0.35 to 0.99. Further validation is required for porosities below 0.35. Nevertheless, the covered porosity range is the relevant region for practical applications, since porosities close to the percolation threshold rarely occur in textiles [55].

To verify the fitting and to ensure that the optimization algorithm did not reach a local minimum when fitting the parameters, the fitted correlation was compared to the geometrical mean of the simulation data. The comparison is shown in Figure 10.

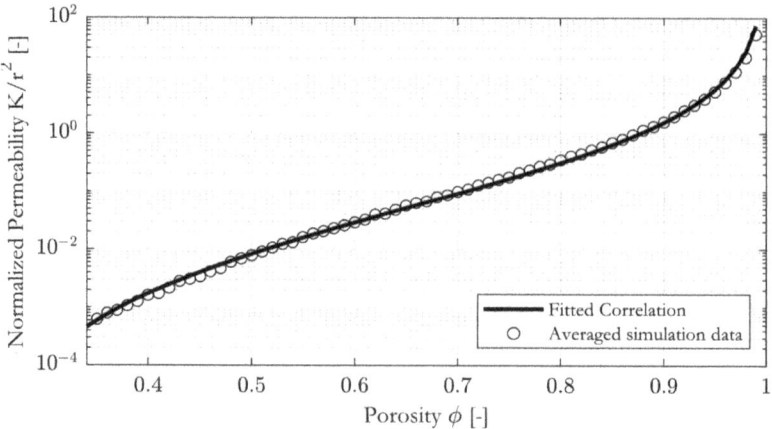

Figure 10. Constitutive permeability–porosity correlation with geometrical mean of simulation data.

The fitted correlation almost perfectly crosses the averaged data points, and $R^2 = 0.998$ is obtained.

Direct comparison of the modified correlation with that by Gebart [8] for hexagonal and squared lattices shows the impact of the random arrangement of fibers for the predicted permeability. Figure 11 shows the comparison between the random, squared and hexagonal arrangement of the fibers. Due to the logarithmic scaling of the y-axes, the differences are quite large even though the curves are relatively close to each other.

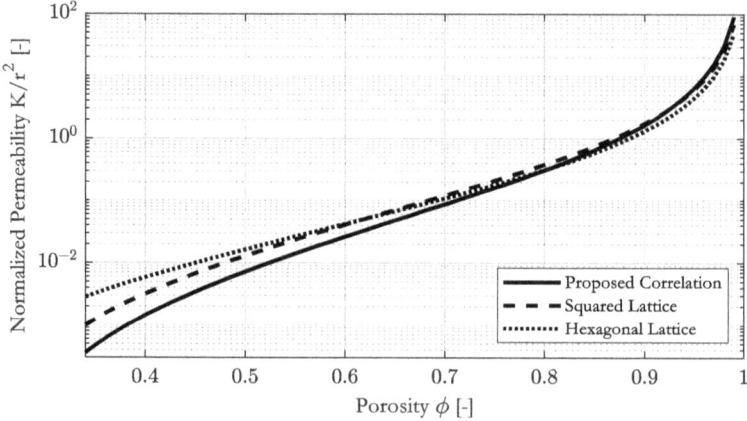

Figure 11. Comparison of the proposed correlations to correlations by Gebart [8] for a squared and a hexagonal arrangement of the fibers, to estimate the permeability of fibrous media.

The correlations of the ordered-arranged fibers predict lower permeability than the correlation for randomly ordered fibers below 0.8. This might explain why Gebart overestimated the permeability when compared to experimental findings for a squared arrangement.

3.3. Constitutive Diffusivity–Porosity Relationship for Randomly Arranged Parallel Fibers

Similar to permeability, we determined the effective diffusion coefficient of 300 random RVEs in the x- and y-direction for each porosity. The porosity of the RVE was varied from 0.35 to 0.99. Again, 38,400 data points were calculated. The simulation results are plotted in Figure 12. As expected, the effective diffusion coefficient tends towards 1 in the limit $\phi \to 1$ and drops towards zero at low porosity. As already observed for the permeability, the variation in the data is greatest at low porosities, where the random arrangement of the fibers leads to large relative changes in the predicted permeability. We found that a modified version of the Perrins [28] relationship provided an excellent fit to the data across the entire porosity range. We adapt Perrin's original relationship by adapting two parameters.

$$D_{eff} = \frac{1}{\phi}\left(1 - \frac{2(1-\phi)}{2 - \phi - C_1(1-\phi)^{C_2}}\right), \quad (19)$$

where C_1 and C_2 are related to the geometry of the RVE.

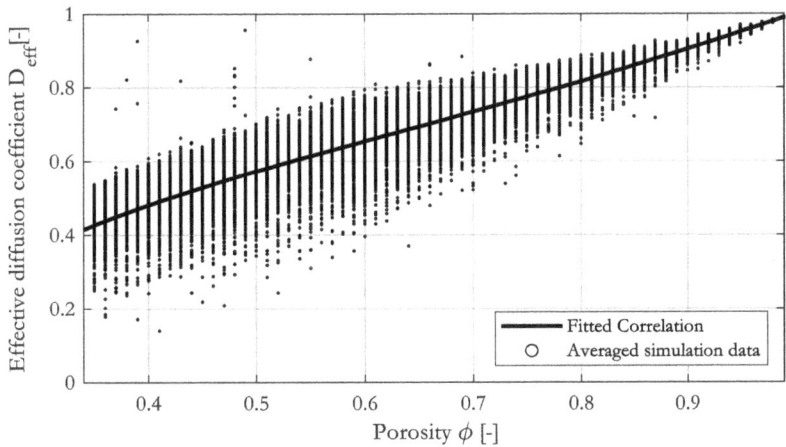

Figure 12. Constitutive diffusion–porosity correlation and simulation data.

Again, we used the lsqcurvefit function of Matlab®, with a Levenberg–Marquardt algorithm for minimizing the Euclidean norm. The fitted parameters are listed in Table 2.

Table 2. Fit values of the diffusion–porosity correlation.

Parameter	Best Fit Value
C_1	0.1711
C_2	0.7895

To validate the fit, we compared the correlation to the geometrical mean of the simulation data in Figure 13. As it was observed for the permeability, the fitted correlation runs almost perfectly through the averaged data, giving $R^2 = 0.999$.

Figure 14 shows a comparison of our proposed correlation to the correlations by Perrins [28] for fibers in a squared and a hexagonal arrangement.

The difference between the correlations is obvious. Especially for a porosity below 0.8, the difference becomes significant. Effective diffusion coefficients calculated from ordered arrangements of fibers are much larger compared to random arrangements.

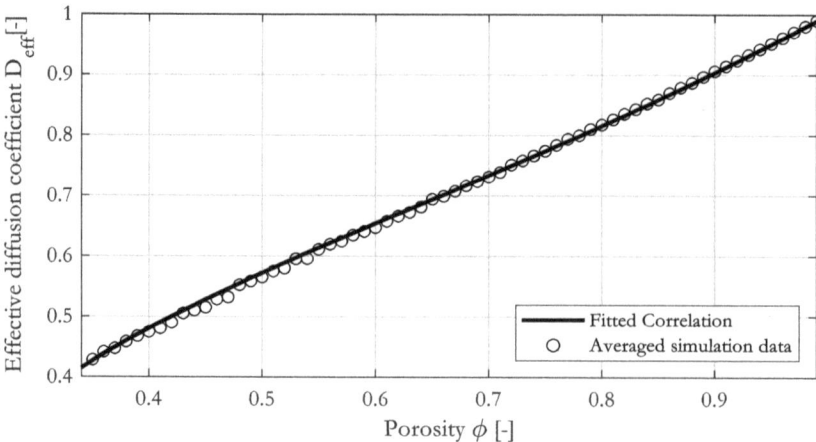

Figure 13. Constitutive diffusion–porosity correlation with geometrical mean of simulation data.

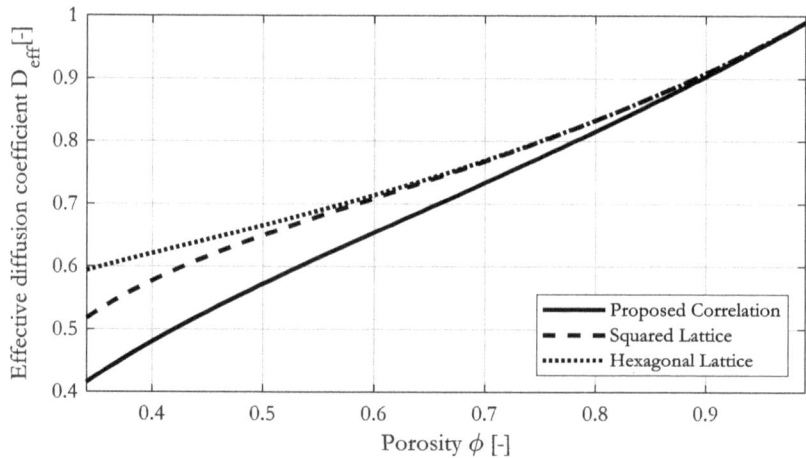

Figure 14. Comparison of the proposed correlation to correlations by Perrins et al. [28] for fibrous media to estimate the effective diffusion coefficient.

4. Conclusions

In this work, we present improved constitutive transport correlations for diffusive and convective mass transport in yarns. The proposed correlations were determined by a new approach using digital reconstruction of the yarn and asymptotic homogenization to work out the transport parameters. We propose to generate a large number of RVEs to statistically represent the microstructure of the porous material under consideration. The cell problems arising from the asymptotic expansion are then solved on the RVEs. The constitutive transport correlations are obtained by curve fitting to calculate the transport parameters. The proposed method for deriving constitutive transport correlations can be applied in future studies to other manufactured porous materials for industrial applications, e.g., battery electrodes, catalysts and filters. The prerequisites are a clear scale separation and a porous structure that can be represented in an RVE and modified by varying the characteristic structural properties, e.g., particle-size distribution or porosity.

The newly derived correlations for yarns facilitate a more accurate prediction of the transport across the fiber banks in the yarns and tows and give a more detailed description of the transport phenomena within the structures. We adapted the correlations proposed by Gebart and Perrins to a more realistic geometric representation of real textile structures [8,28]. We compared the proposed correlations with those in the literature, where fibers are arranged in a squared or hexagonal pattern. The comparison showed that the random arrangement significantly affects transport across the fibrous media, which is in agreement with experimental results in resin transfer moulding [8]. However, the proposed correlations are only applicable to yarns with parallel oriented fibers with a circular cross-section and in a porosity range from 0.35 to 0.99. Additionally, the permeability correlation is only applicable when the flow in the yarn is in the Stokes regime (Reynolds Number $\ll 1$).

In the future, we will compare the proposed correlations with experimental fabric transport measurements to investigate the influence of the yarn structure on the transport properties of the fabric.

Author Contributions: Conceptualization, L.M.; methodology, L.M.; software, L.M. and L.K.-S.; resources, L.P.; writing—original draft preparation, L.M.; writing—review and editing, L.P., M.H.-H. and U.N.; supervision, U.N.; project administration, L.M.; funding acquisition, G.T.G. and U.N. All authors have read and agreed to the published version of the manuscript.

Funding: The authors gratefully acknowledge the project funding of the German Research Council (DFG)—Project Number 453311482. This publication was funded by the German Research Foundation (DFG) grant "Open Access Publication Funding/2023-2024/University of Stuttgart" (512689491).

Institutional Review Board Statement: Not applicable.

Informed Consent Statement: Not applicable.

Data Availability Statement: The data and results involved in this study have been presented in detail in the paper.

Conflicts of Interest: The authors declare no conflict of interest.

References

1. Bartels, V.T. Physiological comfort of sportswear. In *Textiles in Sport*; Woodhead Publishing: Cambridge, UK, 2005; pp. 177–203. [CrossRef]
2. Hasanpour, S.; Ahadi, M.; Bahrami, M.; Djilali, N.; Akbari, M. Woven gas diffusion layers for polymer electrolyte membrane fuel cells: Liquid water transport and conductivity trade-offs. *J. Power Sources* **2018**, *403*, 192–198. [CrossRef]
3. Maduna, L.; Patnaik, A. Heat, moisture and air transport through clothing textiles. *Text. Prog.* **2021**, *52*, 129–166. [CrossRef]
4. Csoklich, C.; Schmidt, T.J.; Büchi, F.N. High performance gas diffusion layers with added deterministic structures. *Energy Environ. Sci.* **2022**, *15*, 1293–1306. [CrossRef]
5. Griebel, M.; Klitz, M. Homogenization and numerical simulation of flow in geometries with textile microstructures. *Multiscale Model. Simul.* **2010**, *8*, 1439–1460. [CrossRef]
6. Šimáček, P.; Advani, S.G. Permeability model for a woven fabric. *Polym. Compos.* **1996**, *17*, 887–899. [CrossRef]
7. Nedanov, P.B.; Advani, S.G. Numerical computation of the fiber preform permeability tensor by the homogenization method. *Polym. Compos.* **2002**, *23*, 758–770. [CrossRef]
8. Gebart, B.R. Permeability of Unidirectional Reinforcements for RTM. *J. Compos. Mater.* **2016**, *26*, 1100–1133. [CrossRef]
9. Becker, J.; Schulz, V.; Wiegmann, A. Numerical determination of two-phase material parameters of a gas diffusion layer using tomography images. *J. Fuel Cell Sci. Technol.* **2008**, *5*, 021006. [CrossRef]
10. Puszkarz, A.K.; Krucińska, I. Modeling of Air Permeability of Knitted Fabric Using the Computational Fluid Dynamics. *Autex Res. J.* **2018**, *18*, 364–376. [CrossRef]
11. Whitaker, S. Flow in porous media I: A theoretical derivation of Darcy's law. *Transp. Porous Media* **1986**, *1*, 3–25. [CrossRef]
12. Bear, J. Theory and Applications of Transport in Porous Media Modeling Phenomena of Flow and Transport in Porous Media. Available online: http://www.springer.com/series/6612 (accessed on 1 March 2022).
13. Turan, R.B.; Okur, A.; Deveci, R.; Açikel, M. Predicting the intra-yarn porosity by image analysis method. *Text. Res. J.* **2012**, *82*, 1720–1728. [CrossRef]
14. Nabovati, A.; Llewellin, E.W.; Sousa, A.C.M. A general model for the permeability of fibrous porous media based on fluid flow simulations using the lattice Boltzmann method. *Compos. Part A Appl. Sci. Manuf.* **2009**, *40*, 860–869. [CrossRef]
15. Sangani, A.S.; Acrivos, A. International Journal of Multiphase Flow 1982. Slow Flow Past Periodic Arrays of Cylinders with Application to Heat Transfer, Elsevier. Available online: https://www.sciencedirect.com/science/article/pii/0301932282900295 (accessed on 18 July 2022).

16. Drummond, J.E.; Tahir, M. International Journal of Multiphase Flow 1984, Laminar Viscous Flow through Regular Arrays of Parallel Solid Cylinders, Elsevier. Available online: https://www.sciencedirect.com/science/article/pii/030193228490079X (accessed on 18 July 2022).
17. Mason, E.; Malinauskas, A. *Gas Transport in Porous Media: The Dusty-Gas Model*; Elsevier: Amsterdam, The Netherlands, 1983.
18. Clague, D.S.; Kandhai, B.D.; Zhang, R.; Sloot, P.M.A. Hydraulic permeability of (un)bounded fibrous media using the lattice Boltzmann method. *Phys. Rev. E* **2000**, *61*, 616. [CrossRef]
19. Koponen, A.; Kandhai, D.; Hellen, E.; Alava, M.; Hoekstra, A.; Kataja, M.; Timonen, J. Permeability of Three-Dimensional Random Fiber Webs. *Phys. Rev. Lett.* **1998**, *80*, 716. [CrossRef]
20. Schulz, R.; Ray, N.; Zech, S.; Rupp, A.; Knabner, P. Beyond Kozeny–Carman: Predicting the Permeability in Porous Media. *Transp. Porous Media* **2019**, *130*, 487–512. [CrossRef]
21. Ingham, D.B.; Pop, I.I. *Transport Phenomena in Porus Media II*; Elsevier: Amsterdam, The Netherlands, 2002; p. 449.
22. Maier, L.; Scherle, M.; Hopp-Hirschler, M.; Nieken, U. Effective transport parameters of porous media from 2D microstructure images. *Int. J. Heat Mass Transf.* **2021**, *175*, 121371. [CrossRef]
23. Šolcová, O.; Šnajdaufová, H.; Schneider, P. Multicomponent counter-current gas diffusion in porous solids: The Graham's-law diffusion cell. *Chem. Eng. Sci.* **2001**, *56*, 5231–5237. [CrossRef]
24. Jó, H.; Halle, B. Solvent Diffusion in Ordered Macrofluids: A Stochastic Simulation Study of the Obstruction Effect. 1996. Available online: http://jcp.aip.org/jcp/copyright.jsp (accessed on 18 July 2022).
25. Bruna, M.; Chapman, S.J. Diffusion in spatially varying porous media. *Source SIAM J. Appl. Math.* **2015**, *75*, 1648–1674. [CrossRef]
26. Hales, J.D.; Tonks, M.R.; Chockalingam, K.; Perez, D.M.; Novascone, S.R.; Spencer, B.W.; Williamson, R.L. Asymptotic expansion homogenization for multiscale nuclear fuel analysis. *Comput. Mater. Sci.* **2015**, *99*, 290–297. [CrossRef]
27. Transvalidou, F.; Sotirchos, S.V. Effective Diffusion Coefficients in Square Arrays of Filament Bundles. *AIChE J.* **1996**, *42*, 2426–2438. [CrossRef]
28. Perrins, W.T.; McKenzie, D.R.; McPhedran, R.C. Transport properties of regular arrays of cylinders. *Proc. R. Soc. London. A. Math. Phys. Sci.* **1979**, *369*, 207–225. [CrossRef]
29. Shou, D.; Fan, J.; Ding, F. Effective diffusivity of gas diffusion layer in proton exchange membrane fuel cells. *J. Power Sources* **2013**, *225*, 179–186. [CrossRef]
30. Koch, D.L.; Brady, J.F. The effective diffusivity of fibrous media. *AIChE J.* **1986**, *32*, 575–591. [CrossRef]
31. Nilsson, L.; Stenström, S. Gas diffusion through sheets of fibrous porous media. *Chem. Eng. Sci.* **1995**, *50*, 361–371. [CrossRef]
32. Rayleigh, L. LVI. On the influence of obstacles arranged in rectangular order upon the properties of a medium. *Lond. Edinb. Dublin Philos. Mag. J. Sci.* **1892**, *34*, 481–502. [CrossRef]
33. Churakov, S.V.; Gimmi, T. Up-scaling of molecular diffusion coefficients in clays: A two-step approach. *J. Phys. Chem. C* **2011**, *115*, 6703–6714. [CrossRef]
34. Helmig, R.; Niessner, J.; Flemisch, B.; Wolff, M.; Fritz, J. Efficient Modeling of Flow and Transport in Porous Media Using Multiphysics andMultiscale Approaches. In *Handbook of Geomathematics*; Springer: Berlin/Heidelberg, Germany, 2010; pp. 417–457. [CrossRef]
35. Battiato, I.; Ferrero, V.P.T.; O'Malley, D.; Miller, C.T.; Takhar, P.S.; Valdés-Parada, F.J.; Wood, B.D. Theory and Applications of Macroscale Models in Porous Media. *Transp. Porous Media* **2019**, *130*, 5–76. [CrossRef]
36. Peyrega, C.; Jeulin, D. Estimation of acoustic properties and of the representative volume element of random fibrous media. *J. Appl. Phys.* **2013**, *113*, 104901. [CrossRef]
37. Kamiński, M.; Lauke, B. Uncertainty in effective elastic properties of particle filled polymers by the Monte-Carlo simulation. *Compos. Struct.* **2015**, *123*, 374–382. [CrossRef]
38. Bensoussan, A.; Lions, J.L.; Papanicolaou, G.C. Boundary Layers and Homogenization of Transport Processes. *Publ. Res. Inst. Math. Sci.* **1979**, *15*, 53–157. [CrossRef]
39. Allaire, G. Homogenization of the stokes flow in a connected porous medium. *Asymptot. Anal.* **1989**, *2*, 203–222. [CrossRef]
40. Auriault, J.-L.; Sanchez-Palencia, E. A Study of the Macroscopic Behavior of a Deformable Saturated Porous Medium. *J. Mécanique* **1977**, *16*, 575–603.
41. Auriault, J.L. Heterogeneous medium. Is an equivalent macroscopic description possible? *Int. J. Eng. Sci.* **1991**, *29*, 785–795. [CrossRef]
42. Pinder, G.F.; Gray, W.G. *Essentials of Multiphase Flow and Transport in Porous Media*; Wiley: Hoboken, NJ, USA, 2008.
43. Matthies, J.H.; Hopp-Hirschler, M.; Uebele, S.; Schiestel, T.; Osenberg, M.; Manke, I.; Nieken, U. Up-scaling transport in porous polymer membranes using asymptotic homogenization. *Int. J. Numer. Methods Heat Fluid* **2020**, *30*, 266–289. [CrossRef]
44. Davarzani, H.; Marcoux, M.; Costeseque, P.; Quintard, M. Experimental measurement of the effective diffusion and thermodiffusion coefficients for binary gas mixture in porous media. *Chem. Eng. Sci.* **2010**, *65*, 5092–5104. [CrossRef]
45. Korneev, S.; Arunachalam, H.; Onori, S.; Battiato, I. A Data-Driven Multiscale Framework to Estimate Effective Properties of Lithium-Ion Batteries from Microstructure Images. *Transp. Porous Media* **2020**, *134*, 173–194. [CrossRef]
46. Zarandi, M.A.F.; Pillai, K.M.; Kimmel, A.S. Spontaneous imbibition of liquids in glass-fiber wicks. Part I: Usefulness of a sharp-front approach. *AIChE J.* **2018**, *64*, 294–305. [CrossRef]
47. Zarandi, M.A.F.; Pillai, K.M. Spontaneous imbibition of liquid in glass fiber wicks, Part II: Validation of a diffuse-front model. *AIChE J.* **2018**, *64*, 306–315. [CrossRef]
48. Matsumoto, M.; Nishimura, T. Mersenne twister. *ACM Trans. Model. Comput. Simul.* **1998**, *8*, 3–30. [CrossRef]

49. Hornung, U. *Homogenization and Porous Media*; Springer: New York, NY, USA, 1997.
50. Alappat, C.; Basermann, A.; Bishop, A.R.; Fehske, H.; Hager, G.; Schenk, O.; Wellein, G. A Recursive Algebraic Coloring Technique for Hardware-efficient Symmetric Sparse Matrix-vector Multiplication. *ACM Trans. Parallel Comput.* **2020**, *7*, 1–37. [CrossRef]
51. *Non-Homogeneous Media and Vibration Theory*; Springer: Berlin/Heidelberg, Germany, 1980.
52. Papanicolau, G.; Bensoussan, A.; Lions, J. *Asymptotic Analysis for Periodic Structures*; AMS: Haarlemmermeer, The Netherlands, 1978.
53. Bakhvalov, N.S.; Panasenko, G.P. *Homogenisation: Averaging Processes in Periodic Media: Mathematical Problems in the Mechanics of Composite Materials*; Springer Science: Berlin/Heidelberg, Germany, 1989.
54. King, P.R. The use of renormalization for calculating effective permeability. *Transp. Porous Media* **1989**, *4*, 37–58. [CrossRef]
55. Chen, X. *Modelling and Predicting Textile Behaviour*; Woodhead Publishing Ltd.: Cambridge, UK, 2009.

Disclaimer/Publisher's Note: The statements, opinions and data contained in all publications are solely those of the individual author(s) and contributor(s) and not of MDPI and/or the editor(s). MDPI and/or the editor(s) disclaim responsibility for any injury to people or property resulting from any ideas, methods, instructions or products referred to in the content.

Article

Free Convection and Heat Transfer in Porous Ground Massif during Ground Heat Exchanger Operation

Borys Basok [1], Borys Davydenko [1], Hanna Koshlak [2,*] and Volodymyr Novikov [1]

[1] Department of Thermophysical Basics of Energy-Saving Technologies, Institute of Engineering Thermophysics National Academy of Sciences of Ukraine, 03057 Kyiv, Ukraine; basok@ittf.kiev.ua (B.B.); bdavydenko@ukr.net (B.D.); nvg52@i.ua (V.N.)
[2] Department of Building Physics and Renewable Energy, Kielce University of Technology, 25-314 Kielce, Poland
* Correspondence: hkoshlak@tu.kielce.pl

Abstract: Heat pumps are the ideal solution for powering new passive and low-energy buildings, as geothermal resources provide buildings with heat and electricity almost continuously throughout the year. Among geothermal technologies, heat pump systems with vertical well heat exchangers have been recognized as one of the most energy-efficient solutions for space heating and cooling in residential and commercial buildings. A large number of scientific studies have been devoted to the study of heat transfer in and around the ground heat exchanger. The vast majority of them were performed by numerical simulation of heat transfer processes in the soil massif–heat pump system. To analyze the efficiency of a ground heat exchanger, it is fundamentally important to take into account the main factors that can affect heat transfer processes in the soil and the external environment of vertical ground heat exchangers. In this work, numerical simulation methods were used to describe a mathematical model of heat transfer processes in a porous soil massif and a U-shaped vertical heat exchanger. The purpose of these studies is to determine the influence of the filtration properties of the soil as a porous medium on the performance characteristics of soil heat exchangers. To study these problems, numerical modeling of hydrodynamic processes and heat transfer in a soil massif was performed under the condition that the pores were filled only with liquid. The influence of the filtration properties of the soil as a porous medium on the characteristics of the operation of a soil heat exchanger was studied. The dependence of the energy characteristics of the operation of a soil heat exchanger and a heat pump on a medium with which the pores are filled, as well as on the porosity of the soil and the size of its particles, was determined.

Keywords: porous medium; soil; ground heat exchanger; filtration; heat pump; numerical simulation

Citation: Basok, B.; Davydenko, B.; Koshlak, H.; Novikov, V. Free Convection and Heat Transfer in Porous Ground Massif during Ground Heat Exchanger Operation. *Materials* **2022**, *15*, 4843. https://doi.org/10.3390/ma15144843

Academic Editor: Dominik Brühwiler

Received: 3 June 2022
Accepted: 9 July 2022
Published: 12 July 2022

Publisher's Note: MDPI stays neutral with regard to jurisdictional claims in published maps and institutional affiliations.

Copyright: © 2022 by the authors. Licensee MDPI, Basel, Switzerland. This article is an open access article distributed under the terms and conditions of the Creative Commons Attribution (CC BY) license (https://creativecommons.org/licenses/by/4.0/).

1. Introduction

Today, the active use of renewable energy sources in the energy supply systems of buildings is the main component of energy-efficient and passive construction projects. Among geothermal technologies, the use of a heat pump system with vertical borehole heat exchangers has become the focus of many researchers. Ground source heat pump systems use the stable ground temperature as a source or sink of heat, which is higher and lower than the air temperature in winter and summer, respectively, and therefore, such technologies can provide better thermal efficiency than normal air source heat pumps [1]. However, there are some unsolved problems in ground source heat pump technology. Several reasons for this can be named. First, the problems of coordinating the implementation of the thermodynamic cycle of a heat pump under conditions of constant load on the ground heat exchanger and a variable load on the heat supply system have not been fully solved. Secondly, when designing a geothermal heat pump and determining the dimensions of a ground heat exchanger, one of the problems is the difficulty in determining in detail the thermal characteristics of the ground medium, which are influenced by several factors and, above all, climatic conditions. The authors of [2] concluded that the simplifying assumptions for analytical calculations of the soil temperature distribution are unrealistic, since the change in the properties of the earth's surface by months of the year has a significant impact on the temperature distribution. Bloom et al. reviewed the technical and economic factors affecting the design and performance of vertical geothermal heat pump systems, and assessed the spatial correlation of these factors with geographic components such as geology and climatic conditions. According to their research so far, subsurface characteristics are not adequately considered during the planning and design of small-scale GSHP systems, which causes under- or over sizing and, therefore, a long-term impact on the maintenance costs and payback time of such systems [3]. Several factors, such as the amount of precipitation, air temperature, windiness, and insolation, determine the potential ability of a low-potential heat source to accumulate energy. Other factors affecting the intensity of absorption and accumulation of energy are the composition and properties of the upper soil layer, as well as its structure and moisture. It was noted in [4,5] that the average monthly air temperature and seasonality affect the temperature of soil layers at a depth of more than 1 m, but the soil temperature remains almost constant at a depth of more than 10 m [6]. However, vertical heat exchangers have a much greater installation depth, at which changes in the heat consumption of the soil are also possible, for example, in the presence of convective groundwater flows. In this case, one cannot simply enter the heat amount values recommended for calculations that can be obtained from the soil, since the convective flow significantly increases this potential (for example, up to 140 W/m^2 with an average recommended value of 50 W/m^2). Thus, in conditions of absolute instability of the parameters that the thermodynamic cycle (Hampson–Linde cycle) must provide, the change in the thermodynamic parameters of the heat pump cycle cannot always be compensated by the controls provided for the heat pump. When setting the efficiency parameters of the heat pump system, for example, COP = 5.6, changing the value of low-potential energy can significantly reduce this indicator.

Theoretically, it is believed that an increase in the value of low-potential energy input contributes to an increase in COP. In practice, when reducing the heat demand in a house, one has to use various approaches to reduce the temperature of the superheated thermodynamic carrier, for example, by injecting cold vapor behind the condenser, into the compressor (if the heat pump is equipped with this technology), which reduces COP by up to 30%, i.e., to 3.9 instead of 6. Therefore, when designing heat pump systems and vertical ground heat exchangers, it is important to take into account the thermal properties of the soil and the characteristics of heat transfer in the ground, for example, convective flux in the presence of groundwater, and to perform design calculations of the ground heat exchanger in more detail.

2. Analysis of Research Results and Publications

It is a well-known fact that the U-tube vertical ground heat exchanger is a simple and reliable design [7]. The problem of heat transfer intensification between vertical ground heat exchangers and soil has been studied by numerous researchers [8–10] and, first of all, by numerical modeling [11–13] of heat transfer processes in the soil mass during the operation of a heat pump. When modeling in calculation modules, heat transfer is usually divided into heat transfer inside the ground heat exchanger and heat transfer outside it. To analyze the efficiency of the ground heat exchanger, it is important to take into account the main factors that can affect the heat transfer processes in the heat exchanger and in the environment (soil). These factors include the composition of the cement mortar material [14], layout of the tubes [15], recovery time, depth and diameter of the well [16,17], and soil properties [18]. In case of water flow in the soil, the heat transfer from the circulating fluid to the ground heat exchanger tube can be considered predominantly convective. In the area from the tube of the heat exchanger to the soil, conductive heat transfer prevails.

In [19], the results of computer simulation of the dynamics of the processes of accumulation and extraction of heat in a soil mass with a single vertical heat exchanger are presented. The calculation results are compared with analytical and computational models. Satisfactory agreement between the results is obtained. Study [20] numerically investigates the process of heat transfer in a soil mass containing a horizontal ground heat collector. Temperature conditions for operation are determined for a working ground heat exchanger on the basis of calculations. The model of heat transfer in vertical ground heat exchangers for heat pump systems is also considered in [21]. The solution to the problem of heat transfer in the soil mass is obtained in an explicit form. The resulting mathematical formula satisfactorily describes the temperature regime of the heat exchanger and can be included in software used for the thermal analysis of ground heat exchangers. It also presents a model of heat transfer inside the borehole, taking into account the thermal interaction between the supports of the U-tube. This can be used to build a formula for thermal resistance inside the borehole.

In [22], a three-dimensional model of heat transfer in these systems is presented, using the finite volume method for implementation purposes. The proposed model takes into account the interconnected processes of heat transfer in the tubes and the soil, which is located between the tubes. Comparison of the calculation results for this model with experimental data shows that the model provides a sufficiently high accuracy.

In [23], the results of computer simulation of borehole ground heat exchangers used in geothermal heat pump systems are presented. The results are based on a 3D model using the implicit finite difference method in a rectangular coordinate system. Each well is approximated by a square column bound by the radius of the borehole. To solve the system of difference equations, the iteration method is used at each step. A comparison is made of the results of calculations by this method and by the model of a finite linear source. The discrepancies between the results obtained by the two methods increase together with the size of the borehole.

In [24], a three-dimensional computational hydrodynamic model of a ground source heat pump with several energy accumulators is presented. The model is designed to investigate the heating performance of a system under continuous and intermittent operating conditions and evaluate the system's thermal energy recovery and performance indicators. The 3D model is based on hybrid meshes with unstructured and structured types of tetrahedra and hexagons. Satisfactory agreement between the results of CFD modeling and experimental data is achieved. The study demonstrates that the soil temperature in the intermittent operation mode is higher than in the continuous operation mode of the heat pump. It has been established that intermittent operation not only helps to restore soil temperature, but also improves the overall performance of the system.

In the works cited, the soil is considered to be a continuous medium. It is believed that heat transfer occurs only through the thermal conductivity of the soil and depends on the thermal resistance of the heat exchanger [25,26]. In reality, the soil is a porous medium, and its pores can be filled with air and fluids [27–29]. In this regard, in addition to thermal conductivity, heat transfer in the soil can also occur by convection of fluids or gases in a porous medium. Convection can be either natural (due to the presence of a temperature gradient in the mass) or forced (or mixed) in case of a pressure gradient in the soil mass. Models of mass, momentum, and heat transfer through a porous medium are being developed to numerically study this problem. In [30], to assess the effect of groundwater flow on the operation of geothermal heat exchangers in heat pump systems with a ground heat source, the heat transfer equation is applied, taking into account advection in a porous medium, and its analytical solution is obtained based on Green's function method. This method is used to determine the effect of groundwater advection on heat transfer. Calculations show that water advection in a porous medium can significantly change the temperature distribution in the soil mass. The hydraulic and thermal properties of soils and rocks affecting the transfer of heat by advection are described.

In [31], the influence of natural convection on the operation of heat exchangers in closed-loop geothermal systems is studied. For numerical study of this problem the Darcy–Brinkman–Forchheimer model [32,33] is used. The basic equations of continuity, momentum, and energy balance are derived taking into account the porosity of a soil medium completely saturated with fluid. The flow of fluid in the pores occurs due to natural convection. The discretization method on a structured grid is used for solving the basic equations. The performance of the heat exchanger is estimated by the volume of extracted energy and by the temperature of the heat carrier at the outlet of the heat exchanger. The results are evaluated by comparing them with the results of calculations according to known existing heat transfer models applied for heat transfer only by thermal conduction. The effect of natural convection and the filtration characteristics of the soil on the operation parameters of the heat exchange unit is determined.

Models of flow in porous media are used not only in problems of free convection in a soil massif. They are also used in modeling the flow of nanofluids [34], including micropolar ferrofluids [35].

The analysis of studies devoted to the issue of heat transfer in a soil porous medium demonstrates that most of the applied models of ground heat exchangers operation assume that the soil is a continuous medium with known thermophysical properties and that heat transfer in the soil occurs only through heat conduction. Heat transfer models that consider heat transfer only by thermal conductivity are adequate in cases where convection affects the total volume of heat extracted from the soil to a much lesser extent. This takes place when the permeability of the porous medium for gas or fluid is sufficiently small. If the soil permeability is significant, it is necessary to apply heat transfer models that take into account the presence of filtration transfer. These issues should be investigated in more detail to determine the characteristics of fluid flow and heat transfer in a porous soil mass during the operation of ground heat exchangers.

3. Purpose of the Study

The purpose of this study is to determine the influence of the filtration properties of the soil as a porous medium on the performance characteristics of ground heat exchangers. For the computational study of these issues, numerical simulation of hydrodynamics and heat transfer in the soil mass was performed, with the assumption that the pores are filled only with fluid; that is, the case of single-phase filtration of a substance in a soil mass during the operation of a heat pump installation is considered.

4. Statement of the Problem of Heat Transfer in a Soil Mass in the Presence of Filtration Processes

The problem of the temperature state of the soil mass during the operation of the ground heat exchanger is formulated as follows. The process of heat transfer in the computational domain, which has the shape of a rectangular parallelepiped with sides x_{max}, y_{max}, and z_{max}, is considered. This parallelepiped covers a section of the soil mass with a vertical U-tube heat exchanger filled with a fluid heat carrier circulating through it. The scheme of the computational domain is shown in Figure 1.

The values x_{max}, y_{max}, and z_{max} are chosen so that the processes of heat transfer to the ground heat exchanger have a minimal effect on the temperature conditions at the boundaries of the computational domain. Soil is considered to be a porous medium, assuming the pores are filled with water.

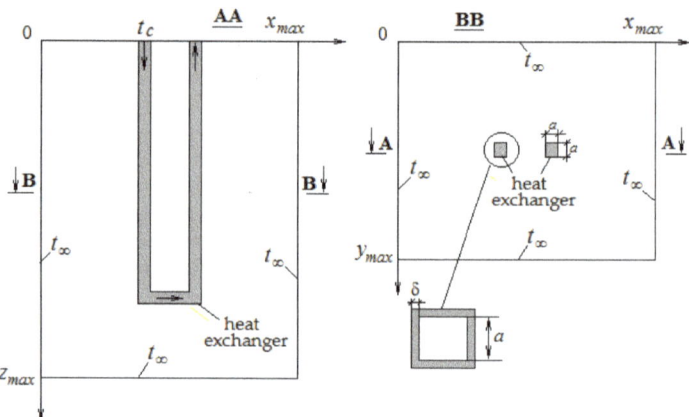

Figure 1. Scheme of the computational domain.

As a result of the difference between the temperature of the heat carrier in the heat exchanger and the temperature of the soil, a free-convection flow of the fluid occurs in the soil mass, between the solid particles of the soil. For simulation of the flow of a fluid in a porous medium, the Darcy–Brinkman–Forchheimer model [32] is used. This flow is described by a system of equations, which includes:

- Continuity equation:

$$\frac{\partial u}{\partial x} + \frac{\partial v}{\partial y} + \frac{\partial w}{\partial z} = 0; \qquad (1)$$

- Momentum equations:

$$\rho_f \left(\frac{1}{\varphi} \frac{\partial u}{\partial \tau} + \frac{u}{\varphi^2} \frac{\partial u}{\partial x} + \frac{v}{\varphi^2} \frac{\partial u}{\partial y} + \frac{w}{\varphi^2} \frac{\partial u}{\partial z} \right) = -\frac{\partial p}{\partial x} + \frac{\mu}{\varphi} \nabla^2 u - \frac{\mu}{K} u - \frac{\rho_f c_F}{\sqrt{K}} |V| u \qquad (2)$$

$$\rho_f \left(\frac{1}{\varphi} \frac{\partial v}{\partial \tau} + \frac{u}{\varphi^2} \frac{\partial v}{\partial x} + \frac{v}{\varphi^2} \frac{\partial v}{\partial y} + \frac{w}{\varphi^2} \frac{\partial v}{\partial z} \right) = -\frac{\partial p}{\partial y} + \frac{\mu}{\varphi} \nabla^2 v - \frac{\mu}{K} v - \frac{\rho_f c_F}{\sqrt{K}} |V| v \qquad (3)$$

$$\rho_f \left(\frac{1}{\varphi} \frac{\partial w}{\partial \tau} + \frac{u}{\varphi^2} \frac{\partial w}{\partial x} + \frac{v}{\varphi^2} \frac{\partial w}{\partial y} + \frac{w}{\varphi^2} \frac{\partial w}{\partial z} \right) = -\frac{\partial p}{\partial z} + \frac{\mu}{\varphi} \nabla^2 w - \frac{\mu}{K} w - \frac{\rho_f c_F}{\sqrt{K}} |V| w - g\beta(t_p - t_\infty) \qquad (4)$$

- Energy equation:

$$C_p \rho_p \left(\frac{\partial t_p}{\partial \tau} + u \frac{\partial t_p}{\partial x} + v \frac{\partial t_p}{\partial y} + w \frac{\partial t_p}{\partial z} \right) = \lambda_p \nabla^2 t_p \qquad (5)$$

where $\nabla^2 = \frac{\partial^2}{\partial x^2} + \frac{\partial^2}{\partial y^2} + \frac{\partial^2}{\partial z^2}$—Laplace operator; $|V|$—fluid flow velocity vector modulus. The vertical coordinate z is directed from the soil surface (z = 0) down. Thermophysical properties of the porous medium are calculated using formulas:

$$\lambda_p = \varphi \lambda_f + (1 - \varphi) \lambda_s;$$

$$C_p \rho_p = \varphi C_f \rho_f + (1 - \varphi) C_s \rho_s.$$

To determine the coefficients K and c_F, the following relationships (6) and (7) are accepted, indicating their dependence on the porosity of the material φ and the diameter of the soil particles d_p

$$K = \frac{d_p^2}{150} \frac{\varphi^3}{(1 - \varphi)^2}; \qquad (6)$$

$$c_F = \frac{1.75}{\varphi^{3/2} 150^{1/2}} \tag{7}$$

The boundary conditions for the system of Equations (1)–(5) are the following:

$$x = 0; \; x = x_{max} : u = 0; \; \frac{\partial v}{\partial x} = 0; \; \frac{\partial w}{\partial x} = 0; t = t_\infty;$$

$$y = 0; \; y = y_{max} : v = 0; \; \frac{\partial u}{\partial y} = 0; \; \frac{\partial w}{\partial y} = 0; t = t_\infty;$$

$$z = 0 : w = 0; \; \frac{\partial u}{\partial z} = 0; \; \frac{\partial v}{\partial z} = 0; \; \frac{\partial t}{\partial z} = 0;$$

$$z = y_{max} : w = 0; \; \frac{\partial u}{\partial z} = 0; \; \frac{\partial v}{\partial z} = 0; t = t_\infty.$$

As follows from the given boundary conditions for $z = 0$, heat transfer from the outer surface of the soil is not taken into account. To simplify the problem, the heat exchanger channel is represented by two straight vertical sections connected by a horizontal section. Sections of the channel are considered to be square. The length of the sides of the square is a, and the thickness of the channel walls is δ.

For the heat carrier flow in the heat exchanger channel, the energy equation has the following form:

- For the section of the downward flow in the vertical section of the channel:

$$C_c \rho_c \left(\frac{\partial t_c}{\partial \tau} + U \frac{\partial t_c}{\partial z} \right) = \lambda_c \nabla^2 t_c; \tag{8}$$

- For the upstream section in the vertical section of the channel:

$$C_c \rho_c \left(\frac{\partial t_c}{\partial \tau} - U \frac{\partial t_c}{\partial z} \right) = \lambda_c \nabla^2 t_c; \tag{9}$$

- For the horizontal section of the channel:

$$C_c \rho_c \left(\frac{\partial t_c}{\partial \tau} + U \frac{\partial t_c}{\partial x} \right) = \lambda_c \nabla^2 t_c; \tag{10}$$

where $U = \frac{G}{a^2}$.

On the outer surface of the heat exchanger channel in contact with the soil, the following conditions are assumed:

$$u = 0; \; v = 0; \; w = 0; \; -\lambda_p \frac{\partial t_p}{\partial n} = \frac{t_p - t_c}{\frac{\delta}{\lambda_w} + \frac{1}{\alpha_c}}, \tag{11}$$

where n is the direction of the outer normal to the outer surface of the channel; α_c is the heat transfer coefficient in the heat exchanger channel, determined by the formula given in [31]: $\alpha_c = 3.66 \cdot \lambda_c / D_e$; D_e is the equivalent diameter of a square channel.

The system of Equations (1)–(10) with the given boundary conditions is solved using the finite difference method. To solve the system of difference equations of fluid dynamics in a porous medium (1)–(4), the SIMPLE algorithm [32] is used. To solve the energy Equation (5) for a porous medium, together with Equations (8)–(10) for the heat carrier and the conjugation condition (11), an explicit time scheme is used.

5. Results of Numerical Studies and Their Analysis

As an example, a ground heat exchanger is considered, the cross-section of which is a square with the side $a = 0.1$ m. The thickness of the channel walls is 2 mm. Its total length is $L = 32.67$ m.

The material of the heat exchanger channel is polyethylene. Heat carrier consumption is $G = 0.21 \times 10^{-3}$ m^3/s. The heat carrier is an aqueous solution of polypropyleneglycol. Its flow velocity is $U = 0.021$ m/s. The time period during which the heat carrier flows from the inlet of the channel to the outlet is $\Delta \tau = L/U = 1556$ s (25.9 min). At the initial moment, the temperature of the soil mass is $t_\infty = 10$ °C. During 45 min, the temperature of the heat carrier at the inlet to the heat exchanger is $t_c = 5$ °C. Further, during the next 15 min, the cooling of the heat carrier in the heat pump stops, and the heat carrier enters the heat exchanger with a temperature of $t_c = 10$ °C. After 15 min, the heat carrier cooling cycle is repeated. Two variants of soil porosity are considered: $\varphi = 0.48$ and $\varphi = 0.40$ with soil particle diameter of $d_p = 0.5$ mm. The pores contain water. Thermal conductivity for solid particles is taken $\lambda_s = 1.5$ W/m/K and for water—$\lambda_f = 0.58$ W/m/K [31]. At the boundaries of the computational domain, the temperature is maintained at the level $t_\infty = 10$ °C.

The results of the calculation of the temperature regime of the ground heat exchanger are presented in the form of fluid velocity fields in the pores and temperature fields. Velocity and temperature fields in the vertical section of the porous soil mass for the case $\varphi = 0.48$; $d_p = 0.5$ mm are presented in Figure 2. The free-convection flow of ground water in the pores arises due to the difference between the temperatures of the heat carrier and the soil mass. As can be seen from the figure, the water flow in the pores near the cooled surface of the heat exchanger channel is directed downward. On the borders of the calculation area, the flow of water in the pores is directed upwards.

To determine the effect of the medium filling the soil pores on the efficiency of the ground heat exchanger, the problem solved for the case of pores filled with water is also solved for the case of pores filled with air. Thermal conductivity for air is $\lambda_f = 0.026$ W/m/K.

The distribution of the vertical velocity w_z horizontally in the direction of the 0X axis along the line intersecting the heat exchanger at a depth of 9.0 m is presented in Figure 3a. The temperature distribution in the soil mass along this line is presented in Figure 3b. Curves 1 refer to the case when the pores are filled with water, and Curves 2 refer to the case when the pores are filled with air. From Figure 3a (Curve 1), it can be seen that the maximum value of 1.6×10^{-6} m/s for velocity w_z is observed near the surfaces of the channel. For the case of filling the pores with air, the maximum velocity at the outer and inner surfaces of the channel is much higher and amounts to $w_z \sim 4.7 \times 10^{-6}$ m/s (Curve 2).

Changes in time of the heat carrier temperature at the inlet to the heat exchanger, which is specified, and the heat carrier temperature at the outlet of the heat exchanger, which is determined from the calculation, are presented in Figure 4. Data refer to the case of filling the pores with water at $\varphi = 0.48$; $d_p = 0.5$ mm.

Curve 1 shows the temperature of the heat carrier at the inlet to the heat exchanger, and Curve 2 shows the temperature of the heat carrier at the outlet of the heat exchanger. As can be seen from these figures, at the beginning of each hour, the temperature of the heat carrier at the inlet to the heat exchanger channel is 5 °C. The temperature at the outlet of the channel rises relative to the temperature at the inlet. Last 15 min. of every hour the temperature of the heat carrier at the inlet to the heat exchanger rises to $t_c = 10$ °C, i.e., the heat carrier during these 15 min. not cooled in the heat pump. In this case, the time period during which the heat carrier that entered the heat exchanger is completely removed from it is $\Delta \tau = 25.9$ min. The period of time during which the temperature of the heat carrier at the inlet to the heat exchanger rises to a temperature of $t_c = 10$ °C is only 15 min. Therefore, the heat exchanger channel simultaneously contains both cooled and uncooled heat carriers. As a result, the maximum values of the heat carrier temperature at the outlet of the heat exchanger lag significantly in time from the maximum temperature values of the heat carrier at the inlet to the heat exchanger.

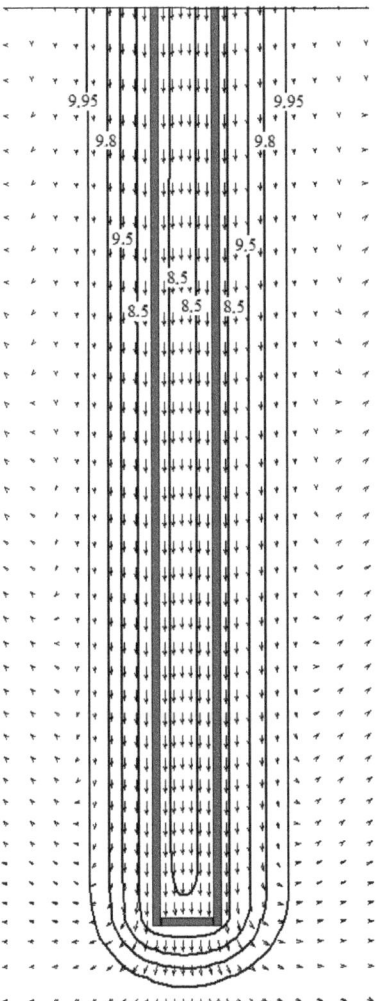

Figure 2. Velocity and temperature distribution (°C) in the soil mass when the pores are filled with water (d_p = 0.5 mm, φ = 0.48). The direction of the vectors coincides with the direction of movement, and the length of the vector is proportional to the fluid velocity.

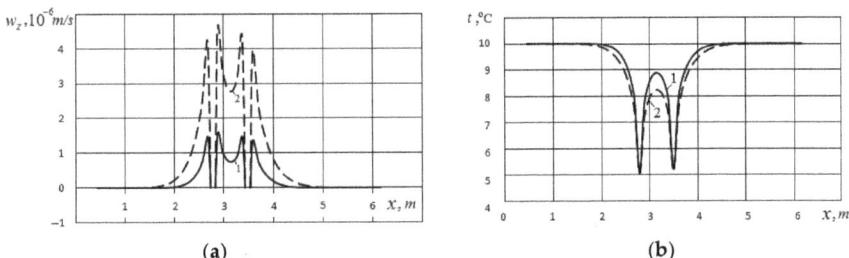

Figure 3. Distribution of vertical velocity (**a**) and temperature (**b**) horizontally at a depth of 9.0 m for the cases of filling pores with water (Curves 1) and with air (Curves 2) at d_p = 0.5 mm; φ = 0.48.

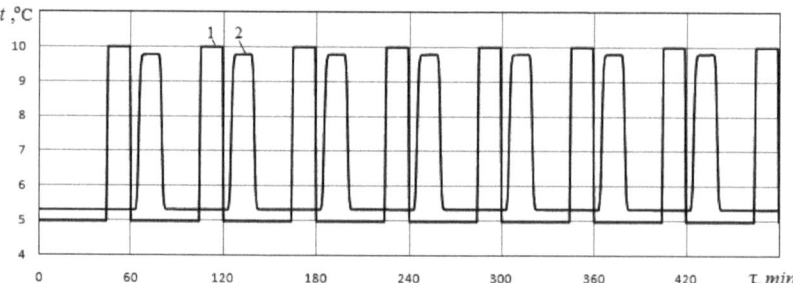

Figure 4. Change in time of the temperature of the heat carrier at the inlet (1) and outlet (2) of the ground heat exchanger for the case of soil porosity $\varphi = 0.48$ ($d_p = 0.5$ mm). The pores are filled with water.

To determine the effect of soil porosity on the energy characteristics of a ground heat exchanger, similar calculations are also performed for the case $\varphi = 0.40$; $d_p = 0.5$ mm. Figure 5 shows the difference between the temperature values at the outlet of the ground heat exchanger for the case of $\varphi = 0.40$ and $\varphi = 0.48$ when the pores are filled with water. This figure shows that the temperature of the heat carrier at the outlet of the heat exchanger channel for the case $\varphi = 0.40$ is $0.002 \ldots 0.016$ °C higher than for the case $\varphi = 0.48$; that is, a greater volume of extracted heat is provided with less porosity.

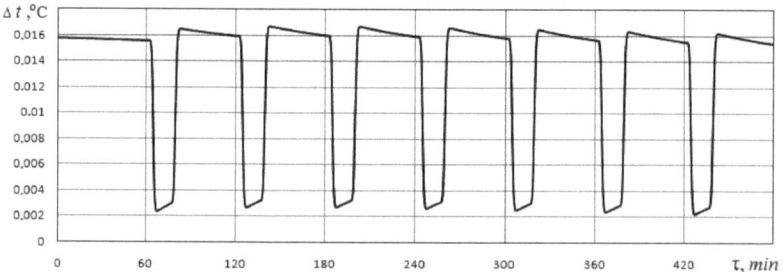

Figure 5. The difference between the temperature values at the outlet of the ground heat exchanger for cases of soil porosity $\varphi = 0.40$ and $\varphi = 0.48$, assuming the pores are filled with water.

The heat carrier temperature distributions along the heat exchanger channel for the cases $\varphi = 0.48$ and $\varphi = 0.40$ at filling the pores with water and air are presented in Figure 6. As can be seen from this figure, for the case of $\varphi = 0.40$, the temperature at the exit from the heat exchanger is somewhat higher than for the case of $\varphi = 0.48$. From this, it follows that a larger volume of extracted heat is provided at a lower porosity. For the case of filling the pores with water, the volumes of heat withdrawn from the soil mass are 219.81 W at $\varphi = 0.48$ and 231.27 W at $\varphi = 0.40$. The difference between these heat volumes is 11.46 W.

Due to the fact that the velocity of free-convection flow of the fluid in the pores under these conditions is very small, the flow of the fluid does not significantly affect the heat transfer from the ground to the heat exchanger. Therefore, the effect of porosity on the intensity of heat transfer is manifested due to the change in the thermophysical properties of the soil with a change in porosity.

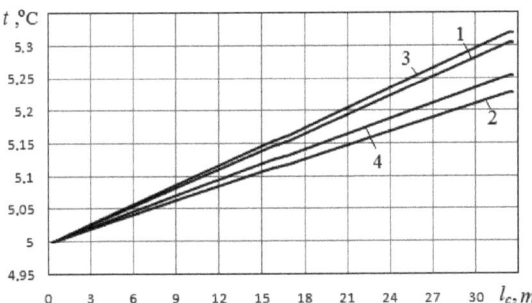

Figure 6. Change in the temperature of the heat carrier along the length of the heat exchanger channel at $d_p = 0.5$ mm for $\varphi = 0.48$ (1, 2) and $\varphi = 0.40$ (3, 4): 1; 3—pores filled with water; 2; 4—pores filled with air.

6. Discussion

As follows from the comparison of the data presented in Figure 3a, referring to the cases of pores filled with water (Curve 1) and with air (Curve 2), the nature of the velocity distribution for the case when the pores are filled with air is qualitatively similar to the case when the pores are filled with water. However, in quantitative terms, the maximum flow velocity when the pores are filled with air is three times higher than in the case of pores filled with water.

As can be seen from the comparison of Curves 1 and 2 in Figure 6, relating to the case $\varphi = 0.48$, for the case of pores filled with water, the temperature at the outlet from the heat exchanger channel is approximately 0.075 °C higher than for the case of pores filled with air. It follows from this that a greater volume of extracted heat is provided when the pores are filled with water. For the conditions under consideration, the volumes of heat extracted from the soil mass are 219.81 W when the pores are filled with water ($d_p = 0.5$ mm; $\varphi = 0.48$) and 165.27 W when the pores are filled with air. The difference between these heat volumes is 54.54 W. For the case $d_p = 0.5$ mm; $\varphi = 0.40$, the difference between the volumes of heat extracted from the soil when the pores are filled with water and air is 48.0 W.

To determine the impact of the medium filling the soil pores on the energy characteristics of the heat exchanger, the difference between the temperature values of the heat carrier at the outlet of the soil heat exchanger for the cases of pores filled with water and filling with air ($d_p = 0.5$ mm; $\varphi = 0.48$) is calculated. The change in time for this difference is presented in Figure 7. Figure 7 shows that the temperature of the heat carrier at the outlet of the heat exchanger in the case of pores filled with water is 0.01 ... 0.08 °C higher than in the case of pores filled with air. This also confirms the fact that the energy characteristics of the heat exchanger are higher when the pores are filled with water compared to when the pores are filled with air.

The novelty of the presented results is that they, in contrast to most published works on the problem of natural convection in soil, were obtained on the basis of solving a three-dimensional problem of free convection for two types of media (water and air) filling soil pores. A porous medium (soil) was represented as solid spherical particles of a given diameter. Two variants of soil porosity were considered. The problem was solved for the area where the vertical ground heat exchanger is located. In fact, the studies were carried out for wet and absolutely dry soil. The physical model is based on the main manifestations of the filtration process—the Darcy, Brinkman, and Forchheimer effects. These calculations were performed for realistic heat pump operation modes—three quarters of the cycle the compressor is working, one quarter is not. The main result of numerical simulation is that, with the same geometry of the heat exchanger, its energy efficiency is higher with less soil porosity. In addition, the heat exchanger works more efficiently in wet soil. This is due to the influence of the thermal conductivity coefficient, the value of which increases with a decrease in soil porosity and in the presence of moisture. But even in absolutely dry soil, it functions quite rationally. Its thermal efficiency is reduced by only 27% compared to wet ground. Due to the low velocity (~10^{-6} m/s) of the free-convective flow of media filling the pores, convection under these conditions has little effect on the heat transfer.

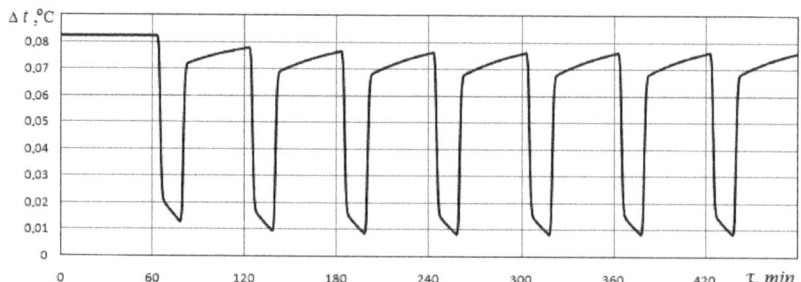

Figure 7. Difference between the values of the heat carrier temperature at the outlet of the ground heat exchanger for the cases of pores filled with water and pores filled with air ($\varphi = 0.48$; $d_p = 0.5$ mm).

The complexity of the practical engineering application of this model lies in the fact that in typical soils with low humidity, the thermal performance of such a single U-tube heat exchanger is not high enough at a depth range of up to 100 m. Therefore, it is advisable to organize a group of such devices to achieve the required thermal power of 0.5–1 MW due to the natural heat of the soil. This raises the problem of optimizing the mutual arrangement of heat exchangers in the group. It is necessary to determine the distances between single heat exchangers, the geometry of their arrangement, and the depth of well drilling. In the problem of the accumulation of heat in the soil and its extraction from the soil, other time intervals arise—seasonal, monthly (ten-day), or daily. These intervals determine the time for the accumulation of heat and for its subsequent extraction.

The proposed numerical model and the results of calculations for this model can be used in the design:

- Systems for low-temperature ground heat extraction for "green" heat pumps;
- Systems of forced accumulation and extraction at the required moment of time of the injected heat, for example, in the conditions of off-season storage of heat or cold;
- Systems that do not use a compressor for air conditioning (for example, a coldwater floor);
- Geothermal ventilation systems;
- Heat and power systems of passive buildings, "zero-energy" buildings;
- Air thermal curtain systems for ventilated building facades.

Some of these problems are already being solved [19,20,33–35].

7. Conclusions

Based on the results of the numerical solution of the system of equations concerning the fluid dynamics and heat transfer in a porous medium filled with water and air, the characteristics of the free-convection fluid flow in a soil mass in the presence of a vertical U-tube ground heat exchanger were obtained. Free-convection flow occurs when there is a temperature gradient in the soil mass, which is a consequence of the operation of the heat pump, one component of which is the ground heat exchanger. The heat pump operates in an intermittent mode, which results in a change in the time of the temperature state of the soil mass and the energy performance of the heat exchanger.

The distributions of temperature and velocity of free-convection flow in a porous soil mass were obtained using computational methods. The results obtained for different values of soil porosity were compared. It was demonstrated that the maximum velocity of free-convection water flow in pores under the considered conditions is of the order of ~10^{-6} m/s. Under these conditions, the effect of natural convection on the heat transfer in the soil mass Is insignificant.

An assessment of the energy performance of the ground heat exchanger, depending on the porosity of the soil, showed that with the remaining operating parameters of the ground heat exchanger being equal, a greater volume of heat extracted from the soil is provided in cases of lower porosity. It also follows from the analysis of the energy characteristics of the heat exchanger operation that these characteristics are higher when the pores are filled with water than when the pores are filled with air; that is, when the pores are filled with water, a greater volume of heat is extracted from the soil compared to the case when the pores are filled with air.

In the future, this numerical model of heat transfer in soil under forced convection of a liquid in a porous medium is supposed to be used in the design of a soil heat exchanger located completely or partially in an aquifer. It is likely that the presence of an aquifer will significantly increase the thermal performance of the heat exchanger. Also of interest is the calculation of heat transfer in the "soil–multipass heat exchanger" system, in which atmospheric air is used as a heat carrier. Such systems are used for geothermal ventilation of the building. Geothermal ventilation is a modern innovative trend in building construction.

Author Contributions: Formal analysis, B.D.; Funding acquisition, H.K.; Methodology, V.N.; Project administration, B.B. and H.K.; Validation, B.B.; Visualization, B.D. and H.K. All authors have read and agreed to the published version of the manuscript.

Funding: This research received no external funding.

Conflicts of Interest: The authors declare no conflict of interest.

Nomenclature

a	Side of the square channel, m
C	Specific heat, J/kg/K
c_F	Forchheimer coefficient
d_p	Soil particle diameter, m
g	Gravity acceleration, m/s^2
G	Heat carrier consumption, m^3/s
K	Permeability, m^2
l_c	Current channel length, m
L	Total channel length, m
p	Pressure, Pa
t	Temperature, °C
u, v, w	Components of the velocity vector, m/s
U	Velocity of heat carrier in the channel, m/s
$x; y; z$	Rectangular coordinates, m
α	Heat transfer coefficient, W/m^2/K
β	Coefficient of thermal expansion, 1/K
δ	Channel wall thickness, mm
λ	Thermal conductivity, W/m/K
μ	Dynamic viscosity, Pa·s
ρ	Density, kg/m^3
τ	Time, s
φ	Porosity

Subscripts:

f	Fluid
p	Porous media
s	Solid
c	Heat carrier
w	Channel wall

References

1. Arif, W.; Youhei, U.; Hikari, F.; Hiroyuki, K.; Isao, T.; Yutaro, S.; Srilert, C.; Punya, C.; Trong, T.T. Numerical simulations on potential application of ground source heat pumps with vertical ground heat exchangers in Bangkok and Hanoi. *Energy Rep.* **2021**, *7*, 6932–6944.
2. Yilmaz, T.; Oezbek, A.; Yilmaz, A.; ve Bueyuekalaca, O. Influence of upper layer properties on the ground temperature distribution. *Isi Bilimi Ve Tek. Derg. J. Therm. Sci. Technol.* **2009**, *29*, 43–51.
3. Eswiasi, A.; Mukhopadhyaya, P. Critical Review on Efficiency of Ground Heat Exchangers in Heat Pump Systems. *Clean Technol.* **2020**, *2*, 204–224. [CrossRef]
4. Fang, L.; Diao, N.; Fang, Z.; Zhu, K.; Zhang, W. Study on the efficiency of single and double U-tube heat exchangers. *Procedia Eng.* **2017**, *205*, 4045–4051. [CrossRef]
5. Liu, X.; Spitler, J.D.; Qu, M.; Shi, L. Recent Developments in the Design of Vertical Borehole Ground Heat Exchangers for Cost Reduction and Thermal Energy Storage. *J. Energy Resour. Technol.* **2021**, *143*, 100803. [CrossRef]
6. Bernier, M. A review of vertical ground heat exchanger sizing tools including an inter-model comparison. *Renew. Sustain. Energy Rev.* **2019**, *110*, 247–265. [CrossRef]
7. Lenhard, R.; Malcho, M. Numerical simulation device for the transport of geothermal heat with forced circulation of media. *Math. Comput. Model.* **2013**, *57*, 111–125. [CrossRef]
8. Omer, A. Heat exchanger technology and applications: Ground source heat pump system for buildings heating and cooling. *MOJ Appl. Bionics Biomech.* **2018**, *2*, 92–107. [CrossRef]
9. Nurullah, K.; Hakan, D. Numerical modelling of transient soil temperature distribution for horizontal ground heat exchanger of ground source heat pump. *Geothermics* **2018**, *73*, 33–47.
10. Sohn, B. Thermal Property Measurement of Bentonite-Based Grouts and Their Effects on Design Length of Vertical Ground Heat Exchanger. *Trans. Korea Soc. Geotherm. Energy Eng.* **2019**, *15*, 1–9.
11. Liu, X.; Xiao, Y.; Inthavong, K.; Tu, J. Experimental and numerical investigation on a new type of heat exchanger in ground source heat pump system. *Energy Effic.* **2015**, *8*, 845–857. [CrossRef]
12. Florides, G.; Kalogirou, S. First in situ determination of the thermal performance of a U-pipe borehole heat exchanger, in Cyprus. *Appl. Therm. Eng.* **2008**, *28*, 157–163. [CrossRef]
13. Luo, J.; Rohn, J.; Bayer, M.; Priess, A. Thermal efficiency comparison of borehole heat exchangers with different drill hole diameters. *Energies* **2013**, *6*, 4187–4206. [CrossRef]
14. Franco, A.; Conti, P. Clearing a Path for Ground Heat Exchange Systems: A Review on Thermal Response Test (TRT) Methods and a Geotechnical Routine Test for Estimating Soil Thermal Properties. *Energies* **2020**, *13*, 2965. [CrossRef]
15. Basok, B.I.; Belyaeva, T.G.; Kuzhel', L.N.; Khibina, M.A. Simulation of the Heat Accumulation and Extraction Processes in the Heat Exchanger–Ground System. *J. Eng. Phys. Thermophys.* **2013**, *86*, 1355–1363. [CrossRef]
16. Basok, B.I.; Davydenko, B.V.; Lunina, A.A. Numerical Model of the Temperature Conditions of the Horizontal Ground Collector. *J. Eng. Phys. Thermophys.* **2012**, *85*, 1114–1126. [CrossRef]
17. Diao, N.R.; Zeng, H.Y.; Fang, Z.H. Improvement in Modeling of Heat Transfer in Vertical Ground Heat Exchangers. *HVACR Res.* **2004**, *10*, 459–470. [CrossRef]
18. Li, Z.; Zheng, M. Development of a Numerical Model for the Simulation of Vertical U-tube Ground Heat Exchangers. *Appl. Therm. Eng.* **2009**, *29*, 920–924. [CrossRef]
19. Lee, C.K.; Lam, H.N. Computer Simulation of Borehole Ground Heat Exchangers for Geothermal Heat Pump Systems. *Renew. Energy* **2008**, *33*, 1286–1296. [CrossRef]
20. Yuanlong, C.; Jie, Z. CFD Assessment of Multiple Energy Piles for Ground Source Heat Pump in Heating Mode. *Appl. Therm. Eng.* **2018**, *139*, 99–112.
21. Basok, B.; Davydenko, B.; Pavlenko, A. Numerical Network Modeling of Heat and Moisture Transfer through Capillary-Porous Building. *Materials* **2021**, *14*, 1819. [CrossRef] [PubMed]
22. Koshlak, H.; Pavlenko, A. Heat and Mass Transfer During Phase Transitions in Liquid Mixtures. *Rocz. Ochr. Srodowiska* **2019**, *21*, 234–249.
23. Hu, W.; Jiang, Y.; Chen, D.; Lin, Y.; Han, Q.; Cui, Y. Impact of Pore Geometry and Water Saturation on Gas Effective Diffusion Coefficient in Soil. *J. Appl. Sci.* **2018**, *8*, 2097. [CrossRef]
24. Chamindu Deepagoda, T.K.K.; Smits, K.; Jayarathne, J.R.R.N.; Wallen, B.M.; Clough, T.J. Characterization of Grain-Size Distribution, Thermal Conductivity and Gas Diffusivity in Variably Saturated Binary Sand Mixtures. *Vadose Zone J.* **2018**, *17*, 1–13. [CrossRef]
25. Pavlenko, A.M.; Koshlak, H. Application of Thermal and Cavitation Effects for Heat and Mass Transfer Process Intensification in Multicomponent Liquid Media. *Energies* **2021**, *14*, 7996. [CrossRef]
26. Diao, N.; Li, Q.; Fang, Z. Heat Transfer in Ground Heat Exchangers with Groundwater Advection. *Int. J. Therm. Sci.* **2004**, *43*, 1203–1211. [CrossRef]
27. Ghoreishi-Madiseh, S.A.; Hassani, F.; Mohammadian, A.; Radziszewski, P. A Transient Natural Convection Heat Transfer Model for Geothermal Borehole Heat Exchangers. *J. Renew. Sustain. Energy* **2013**, *5*, 043104. [CrossRef]
28. Nield, D.A.; Bejan, A. *Convection in Porous Media*; Springer Inc.: New York, NY, USA, 1992.

29. Serageldin, A.A.; Radwan, A.; Sakata, Y.; Katsura, T.; Nagano, K. The Effect of Groundwater Flow on the Thermal Performance of a Novel Borehole Heat Exchanger for Ground Source Heat Pump Systems: Small Scale Experiments and Numerical Simulation. *Energies* **2020**, *13*, 1418. [CrossRef]
30. Alshehri, A.; Shah, Z. Computational Analysis of Viscous Dissipation and Darcy-Forchheimer Porous Medium on Radioactive Hybrid Nanofluid. *Case Stud. Therm. Eng.* **2022**, *30*, 101728. [CrossRef]
31. Shah, Z.; Alzahrani, E.O.; Dawar, A.; Ullah, A.; Khan, I. Influence of Cattaneo-Christov Model on Darcy-Forchheimer Flow of Micropolar Ferrofluid over a Stretching/Shrinking Sheet. *Int. Commun. Heat Mass Transf.* **2020**, *110*, 104385. [CrossRef]
32. Patankar, S.V. *Numerical Heat Transfer and Fluid Flow*; Hemisphere: New York, NY, USA, 1980.
33. Basok, B.; Davydenko, B.; Bozhko, I.; Moroz, M. Unsteady Heat Transfer in a Horizontal Ground Heat Exchanger. *Ind. Heat Eng.* **2018**, *40*, 34–40. [CrossRef]
34. Basok, B.; Davydenko, B.; Bozhko, I.; Nedbailo, A. Three-Dimensional Numerical Model of Hydrodynamics and Heat Transfer in the System Soil–Heat Exchanger–Heat Carrier. In *Book of Abstracts, Actual Problems of Renewable Energy, Construction and Environmental Engineering, Proceedings of the V International Scietific-Technical Conference Actual Problems of Renewable Energy, Construction and Environmental Engineering, Kielce University of Technology, Kielce, Poland, 3–5 June 2021*; Koszalin University of Technology: Koszalin, Poland, 2021; pp. 45–47.
35. Nakorchevskii, A.I. Heat Absorption by a Groundmass Due to the Action of Solar Radiation. *J. Eng. Phys. Thermophys.* **2016**, *89*, 1394–1400. [CrossRef]

Article

Magnetic Induction Assisted Heating Technique in Hydrothermal Zeolite Synthesis

Supak Tontisirin [1,2,*], Chantaraporn Phalakornkule [1], Worawat Sa-ngawong [1] and Supachai Sirisawat [1]

[1] Department of Chemical Engineering, Faculty of Engineering, King Mongkut's University of Technology North Bangkok, Bangkok 10800, Thailand; chantaraporn.p@eng.kmutnb.ac.th (C.P.); worawat.sa@outlook.com (W.S.-n.); supachai.sirisawat@gmail.com (S.S.)

[2] Center of Eco-Materials and Cleaner Technology, King Mongkut's University of Technology North Bangkok, Bangkok 10800, Thailand

* Correspondence: supak.t@eng.kmutnb.ac.th; Tel.: +66-2-555-2000 (ext. 8257 or 8230)

Abstract: The magnetic induction assisted technique is an alternative heating method for hydrothermal zeolite synthesis with a higher heat-transfer rate than that of the conventional convection oil bath technique. The research demonstrates, for the first time, the application of the magnetic induction heating technique with direct surface contact for zeolite synthesis. The magnetic induction enables direct contact between the heat source and the reactor, thereby bypassing the resistance of the heating medium layer. A comparative heat-transfer analysis between the two methods shows the higher heat-transfer rate by the magnetic induction heating technique is due to (1) eight-time higher overall heat-transfer coefficient, attributed to the absence of the resistance of the heating medium layer and (2) the higher temperature difference between the heating source and the zeolite gel. Thereby, this heating technique shows promise for application in the large-scale synthesis of zeolites due to its associated efficient heat transfer. Thus, it can provide more flexibility to the synthesis method under the non-stirred condition, which can create possibilities for the successful large-scale synthesis of a broad range of zeolites.

Keywords: hydrothermal synthesis; zeolite synthesis; magnetic induction; zeolite X; NaX; scaleup

Citation: Tontisirin, S.; Phalakornkule, C.; Sa-ngawong, W.; Sirisawat, S. Magnetic Induction Assisted Heating Technique in Hydrothermal Zeolite Synthesis. *Materials* **2022**, *15*, 689. https://doi.org/10.3390/ma15020689

Academic Editor: Anatoliy Pavlenko

Received: 28 October 2021
Accepted: 12 January 2022
Published: 17 January 2022

Publisher's Note: MDPI stays neutral with regard to jurisdictional claims in published maps and institutional affiliations.

Copyright: © 2022 by the authors. Licensee MDPI, Basel, Switzerland. This article is an open access article distributed under the terms and conditions of the Creative Commons Attribution (CC BY) license (https://creativecommons.org/licenses/by/4.0/).

1. Introduction

Zeolites are crystalline aluminosilicate microporous materials. They can be natural or synthetic [1]. They have a three-dimensional framework structure, which has an open pore with a size of approximately 0.3–1 nm [2]. Due to their high surface area, numerous active sites, high thermal and chemical stability, and shape selectivity, zeolites are widely used in many industrial applications as adsorbents, ion-exchangers, and catalysts [3–7]. Most of their applications are relevant to environmental remediation and renewable energy for a sustainable society [8–12]. Zeolites are the largest volume used as ion-exchangers for reducing the hardness of the water in detergent application. This revolutionary application prevents the death of living creatures in rivers caused by the traditional phosphate substitute compound, via a process known as "Eutrophication". Conversely, the highest market value of zeolites is as catalysts, particularly in refineries because of their unique properties.

Due to their numerous benefits and environmental friendliness, zeolites are demanded in high capacity for various applications. However, their production technique still has room to be improved for the high-quality zeolites at a low production cost. Synthetic zeolites can be produced by hydrothermal method at a high temperature and pressure, in an aqueous solution, and in a closed reactor system [13–15]. In a small scale, the heat sources are typically provided by a convection hot air oven and an oil bath. Since the heat transport mode is mainly convection, the arising problem is the slow heat transfer across the reactor to the core of the zeolite gel, and there could be heat loss from the heat source to the environment. Magnetic induction heating is a suitable technique for directly heating the zeolite gel system. This technique is applied by oscillating the magnetic field, which causes the electrons in the magnetic material to move around and, consequently, creates a large eddy of electron current [16]. In zeolite production by the hydrothermal method, there are two possible methods for heat generation by magnetic induction: (1) via the conductive properties of the solution [17] and (2) via induction of heat in the magnetic vessel or material which contact with the zeolite gel. This can be used by coating application of the stainless-steel substrate immersed in the zeolite gel. The heat in the substrate is generated by induction of the coil by distance [18]. Alternatively, it can be used by direct contact of the magnetic vessel with the reactor surface. For the first method, the heat is generated in the zeolite solution because the gel solution exhibits conductive property (i.e., the ionic strength) due to the mobile ions. However, the effect of magnetic induction on the formation mechanism and destabilization of the zeolite framework is not clearly known. It might destabilize certain crystalline domains of the zeolite, which are crucial for the success of the zeolite synthesis. The second method is more flexible than the first one. That is, it is not limited only to the conductive liquid. It can be applied to all types of liquids. Applying fast heat transfer to the zeolite gel can reduce temperature gradient in the bulk and can reduce the energy supply for the zeolite production. Importantly, the direct contact of the heat from a magnetic vessel with the reactor system or magnetic material contacting with the zeolite gel will not disturb or destabilize the crystalline domains of the zeolite. Therefore, applying the magnetic induction heating by the direct contact of the magnetic vessel to the zeolite is more practical and potential approach for producing different zeolite types.

This research demonstrates, for the first time, the technique of applying magnetic induction heating by direct surface contact of the magnetic vessel with the reactor surface for the bench-scale zeolite synthesis. NaX type zeolite was hydrothermally synthesized under the non-stirred condition. The temperature profiles and product quality at different points in the reactor were investigated to evaluate the heat-transfer performance. The clarification of their advantages by experimental data and theoretical calculation is essential for elucidating this approach. The heating technique by the conventional convection oil bath was conducted for comparison.

2. Materials and Methods

2.1. Apparatus Setup and Synthesis of NaX Zeolite

A 2.5-L turbine-agitator stainless-steel reactor is designed and fabricated. It is placed on the ferromagnetic vessel. The reactor is equipped with a magnetic induction heating plate (max. 1600 W, setting at 600 W), a programmable logic controller (PLC), and a data logger, as shown in Figure 1. The magnetic field is created in this way: the 220 V, 50 Hz alternating current is applied to the induction coil of the magnetic induction heater, creating an alternating electromagnetic field with a frequency of approximately 25 kHz and amplitude in the order of a few micro-tesla (data provided by a commercial supplier). As the alternating magnetic field creates an oscillating electric current and resistance in the ferromagnetic plate, heat is generated inside the metal material by the eddy current. The heat is transferred to the bulk liquid due to the temperature difference between the ferromagnetic plate and the reaction solution. Due to a relatively low magnetic susceptibility of the reaction solution (the CGS volume magnetic susceptibility at 28 °C of -1.752×10^{-6}), direct induction heating of the solution is expected to be negligible. Four temperature sensors at different locations inside the reactor: depth, radius, and angle positions (Ta = 1.5, Tb = 6.5, Tc = 6.5, Td = 11.5 cm from the base of the reactor; Ta = 6.25, Tb = 6.25, Tc = 3.85, Td = 6.25 cm from the center; Ta = 0, Tb = 30, Tc = 30, Td = 90° from the radius of the circle), are installed to monitor their profiles. A pressure transducer is setup for the detection of the pressure inside the reactor. The apparatus is controlled by the PLC, and data are recorded via the data logger. The NaX zeolite is synthesized with the gel composition 4.42 SiO_2: 1 Al_2O_3: 3.71 Na_2O: 539 H_2O. The 7.9 wt.% sodium hydroxide solution (solid NaOH from Merck, 99 wt.%) is added to the sodium aluminate solution (homemade, 10.9 wt.% Al_2O_3,

20.5 wt.% NaOH, 68.6 wt.% H$_2$O) and stirred for 15 min in the beaker. The 7.9 wt.% sodium hydroxide solution is added to sodium silicate solution (Panreac, 27.4 wt.% SiO$_2$, 8.3 wt.% Na$_2$O, 64.3 wt.% H$_2$O) and stirred for 15 min in the beaker. Then, the two solutions are mixed in the reactor with the total gel of ca. 1.5 L. The gel is stirred with the impeller at 290 rpm for 1 h and subjected to static ageing for 16 h. The reaction occurs at 100 °C for 8 h under the non-stirred or static condition. The temperature and pressure inside the reactor are monitored throughout the synthesis. After the reaction is complete, the solid zeolite product at two different points (the rim and center of the reactor) and the rest (bulk) are separately collected. The zeolite is filtered, washed with deionized water, and dried in the oven at 100 °C for 8 h. The phase and crystallinity of the zeolite are determined by X-ray diffraction spectroscopy (XRD). The chemical composition is analyzed by X-ray fluorescence spectroscopy (XRF). The textural properties are investigated by low-temperature N$_2$-adsorption. The analysis of heat transfer using the oil bath technique (2000 W, max. temp. 95 °C) for zeolite synthesis is conducted for comparison with the magnetic induction heating technique. The immersed depth of the reactor in the oil bath is 120 mm.

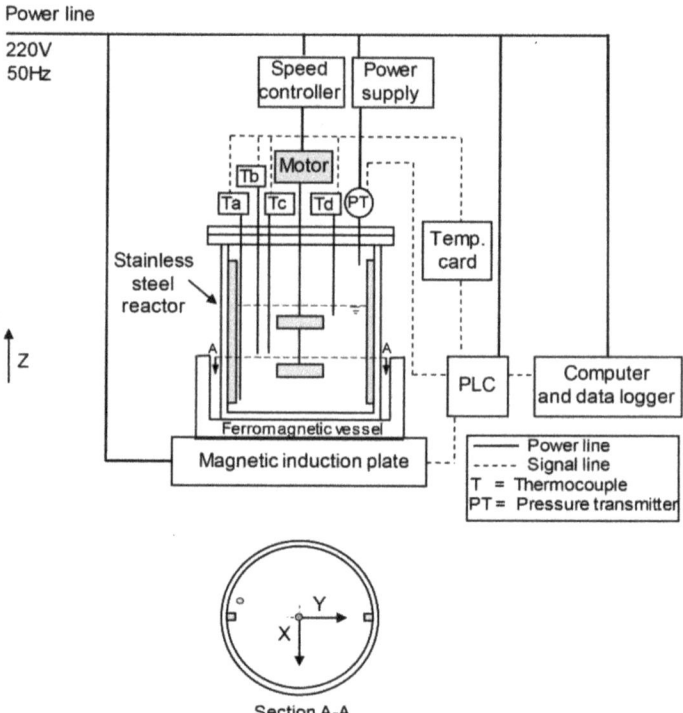

Figure 1. The bench-scale reactor setup with the magnetic induction heating plate.

2.2. Characterization of the NaX Zeolite

The magnetic susceptibility of the reaction solution is measured at 28 °C using a magnetic susceptibility balance (Sherwood Scientific, MSB MK1, Cambridge, UK). The obtained zeolite is characterized by powder X-ray diffraction spectroscopy (XRD, Bruker, AXS model D8 Advance, Karlsruhe, Germany) with CuKa (λ = 1.544 A) operating at 40 kV and 30 mA. The measurement runs with 2 theta from 5 to 50 with a step size of 0.02 and a scan speed of 1 s. The specific surface area and pore volume are determined by low-temperature N$_2$-adsorption at −196 °C (Microtrac MRB, BELSORP-mini, Osaka, Japan). Before determining the adsorption isotherm, the sample is pretreated at 200 °C for 12 h

(BELPREP-flow) to remove the humidity in the zeolite. The specific surface area is calculated by the Brunauer-Emmett-Teller (BET) method in the pressure range of $p/p_0 \approx 0.03$–0.1. The total pore volume is determined by adsorption branch isotherm at $p/p_0 \approx 0.98$. The crystal morphology is revealed using field emission scanning electron microscopy (FESEM, JEOL, JSM-7610, Tokyo, Japan). The elemental composition of the zeolite is measured by X-ray fluorescence spectrometry (XRF, Bruker, S8 TIGER, Karlsruhe, Germany).

3. Results and Discussion

3.1. The Synthesis of the NaX Zeolite

The temperature profiles of the zeolite gel at different locations in the hydrothermal reactor by magnetic induction heating source and the convection oil bath heating source are shown in Figure 2a,b, respectively. In magnetic induction heating, the ferromagnetic vessel is the heating source, which directly conducts heat to the stainless-steel reactor. Thus, the zeolite mixture is rapidly heated. The four temperatures rapidly rise to the target temperature of 100 °C within 17.8 min at a heating rate of 4.2 °C/min. The temperatures at the different locations of Ta, Tb, and Tc (deep side) show similar profiles at average temperature ca. 100 °C, while Td (shallow side) stays rather at a lower temperature of ca. 96 °C. However, its maximum reaches the target temperature. With oil bath heating, the temperatures slowly rise to the target temperature of 95 °C at a heating rate of 1 °C/min. This is due to the slow heat transfer by natural convection of the oil fluid, which necessitates heating the oil body. All temperatures do not achieve the target temperature (Ta is at the highest of 85 °C). The end temperatures are different. This indicates that the heat loss to the environment easily occurs, resulting in inefficient heat transfer. This shows the magnetic induction heating promotes a high heat-transfer rate and minimizes the heat loss. The zeolite products synthesized by two different heating techniques are sampled at different locations in the reactor: rim, center, and the rest of the bulk. In Figure 3, all samples correspond to the FAU framework of NaX zeolite [19]. Figure 3a shows the XRD patterns of NaX by the magnetic induction heating. The similar patterns and intensities of three samples at different locations in the reactor signify that the mass transfer in the zeolite gel is good. This corresponds to efficient heat transfer by homogenous temperature profiles inside the reactor. Differently, the XRD patterns of samples synthesized by oil bath heating present irregular intensities with the lowest one at the center position of the reactor as shown in Figure 3b. This results by inefficient mass and heat transfer from the heat source to the content in the reactor core.

The amount of zeolite product obtained by the magnetic induction heating is 51.6 g, with a 56% yield referring to aluminium element, with a molar ratio of Si/Al = 1.14. By the scanning electron microscopy, the bulk NaX zeolite reveals the intergrowth among octahedral and irregular-shape morphologies with crystal size of ca. 1–1.2 μm as shown in Figure 4a. It possesses specific surface area of 495 m^2/g and pore volume of 0.23 cm^3/g. Whereas, the zeolite synthesized by the oil bath heating is 56.9 g with a 62% yield referring to aluminium, with a molar ratio of Si/Al = 1.22. The crystal morphology is similar to that of the zeolite synthesized by the magnetic induction heating, but its size is smaller ca. 0.8–1 μm (Figure 4b). This shows slightly lower dispersion of the crystal size. Smaller crystal size can contribute to larger contact surface area. The specific surface area and pore volume are 520 m^2/g and 0.24 cm^3/g. The isotherms in both NaX zeolites are presented in Figure 5.

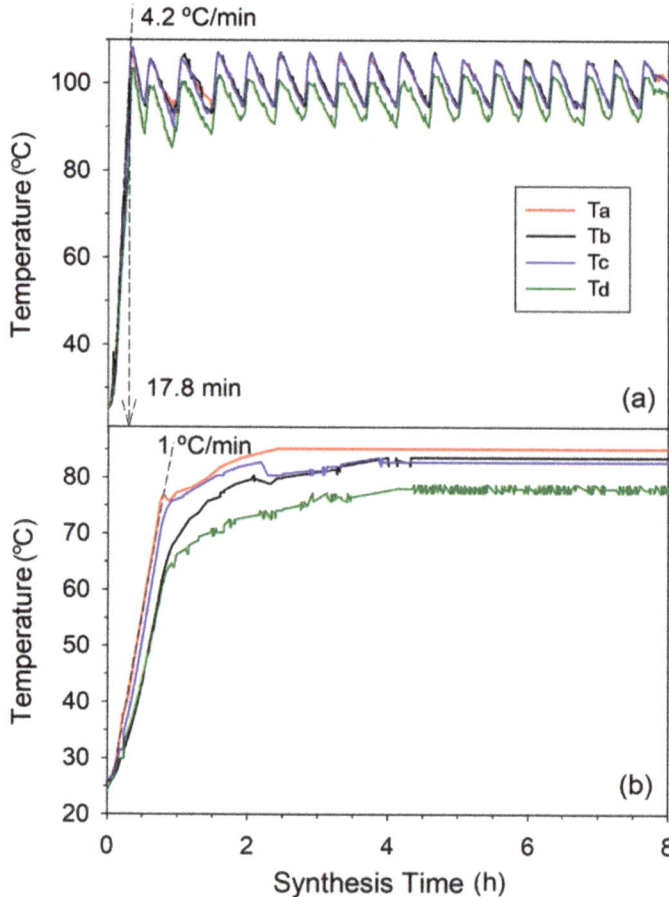

Figure 2. The temperature profiles of the reaction mixture of NaX zeolite for non-stirred synthesis by: (**a**) magnetic induction heating; (**b**) convection oil bath heating.

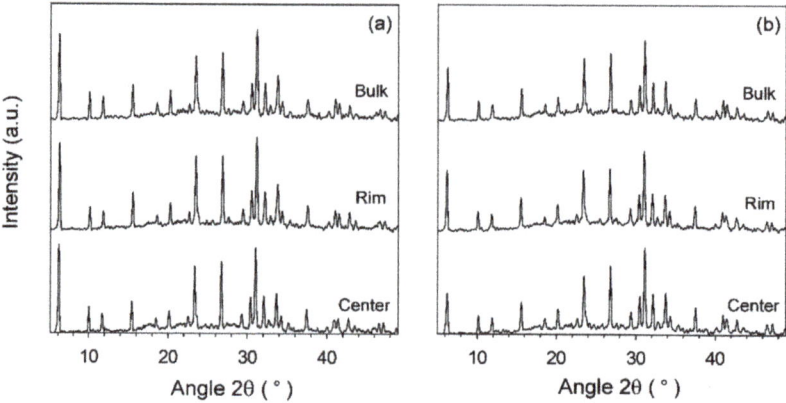

Figure 3. The XRD patterns of NaX zeolite sampled at different locations, synthesized by: (**a**) magnetic induction heating; (**b**) convection oil bath heating.

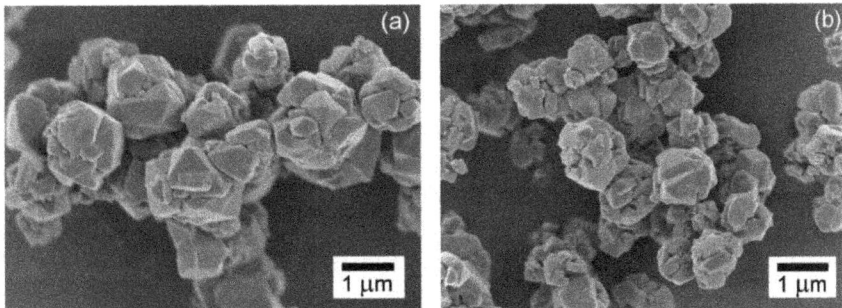

Figure 4. The crystal morphologies of NaX zeolites synthesized by: (**a**) magnetic induction heating; (**b**) convection oil bath heating.

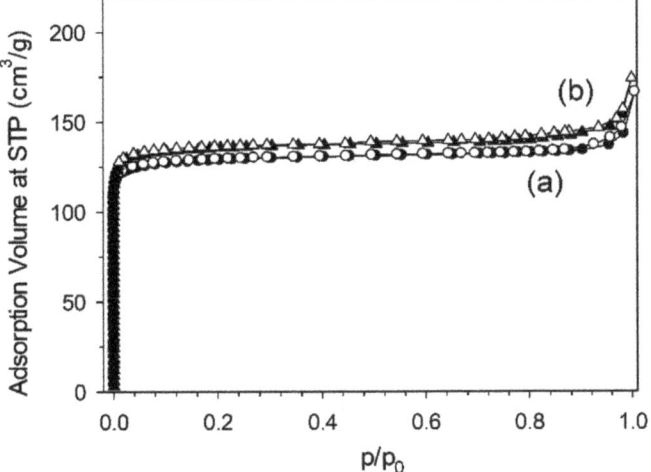

Figure 5. The N_2 adsorption and desorption isotherms of NaX zeolites synthesized by: (**a**) magnetic induction heating; (**b**) convection oil bath heating. (Closed symbol: adsorption; open symbol: desorption).

3.2. Comparative Study of the Heat-Transfer Efficiency Using Magnetic Induction and Convection Oil Bath Heating Techniques

The magnetic induction heating technique is compared with the conventional convection oil bath heating. The geometry of the reactor vessel has a cylindrical shape. However, the cartesian coordinate system is used to describe the space and directions of heat transfer in the reactor vessel because of the symmetry of the properties around the z-axis (independent of the angular positions). The representation of the cartesian coordinate system in the cylindrical vessel is shown in Figure 6. This figure illustrates the temperature gradients from the heat sources to the bulk of zeolite gel in the reactor. In the oil bath heating, the heat transfers in the z and y axis. There are at least three resistances to heat transfer (Figure 6a): Region 0–1 where the reactor bottom and wall is in contact with oil (T_{oil}), and the heat transfer at the boundary is governed by Newton's law of cooling; Region 1–2 in which heat is conducted through the stainless steel and the heat transfer is governed by Fourier's law; Region 2–3 where the reactor bottom and wall is in contact with the zeolite gel at ambient temperature ($T_{zeolite\ gel}$), and the heat transfer at the boundary is governed by Newton's law of cooling.

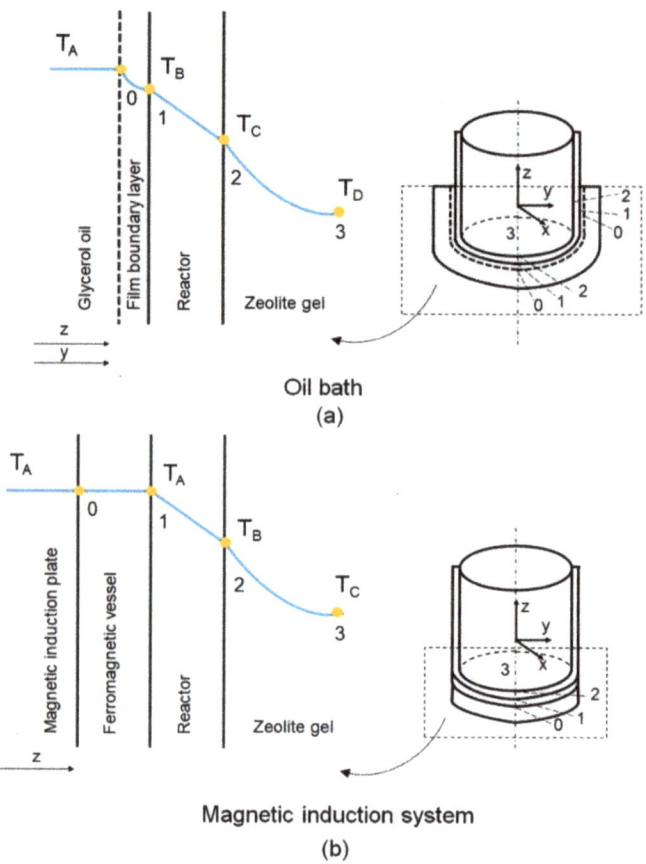

Figure 6. The temperature gradients: (**a**) in the glycerol oil bath heating system; (**b**) in the magnetic induction heating system.

The heat-transfer equation for the convection oil bath heating is given in Equation (1):

$$Q_{oil\ bath} = \frac{\Delta T_{oil\ bath} U_{oil\ bath}}{\frac{1}{A}} = \frac{\Delta T_{oil\ bath}}{\frac{1}{h_1 A} + \frac{L}{kA} + \frac{1}{h_2 A}} \quad (1)$$

where

$Q_{oil\ bath}$ is the amount of heat flow in the oil bath heating system (W)
$\Delta T_{oil\ bath}$ is the temperature difference between T_{oil} and $T_{zeolite\ gel}$, 70 (K)
$U_{oil\ bath}$ is the overall heat-transfer coefficient in the glycerol oil bath heating system (W/m²·K)
A is the heat-transfer surface area of the reactor (m²)
L is the thickness of the reactor, 0.003 (m)
h_1 is the free convective heat-transfer coefficient of the glycerol oil (Region 0–1) (W/m²·K)
h_2 is the free convective heat-transfer coefficient of the zeolite gel (Region 2–3) (W/m²·K)
k is the thermal conductivity of the stainless steel (Region 1–2), 16 (W/m·K)

$U_{oil\ bath}$ is calculated according to Equation (2):

$$U_{oil\ bath} = \frac{1}{\frac{1}{h_1} + \frac{L}{k} + \frac{1}{h_2}} \quad (2)$$

In convection oil bath heating, $U_{oil\ bath}$ is determined from both horizontal plate and vertical plane/cylinder because the glycerol oil encompasses the reactor bottom and side wall as shown in Figure 6a. For the horizontal plate of the reactor bottom, h_1 can be calculated according to the relation of dimensionless groups of Nusselt number (Nu) in Equation (3), when the product of the Grashof number (Gr) and the Prandtl number (Pr) is less than 2×10^8 [20].

$$Nu = 0.13(Gr \cdot Pr)^{1/3} \tag{3}$$

$$h_1 = \frac{k_1}{l} 0.13(Gr \cdot Pr)^{1/3} \tag{4}$$

where

k_1 is the heat conductivity of glycerol oil at reference temperature (W/m·K)
l is the characteristic length of heat-transfer surface (m)

The physical properties are determined at the reference temperature according to Equation (5) [21–23].

$$T_e = T_w - 0.25(T_w - T_\infty) \tag{5}$$

where

T_e is the reference temperature (K)
T_w is the average wall temperature (K)
T_∞ is the free-stream temperature (K)

The Grashof number is calculated according to Equation (6) to be 15,602.

$$Gr = \frac{l^3 \rho^2 g \beta \Delta T}{\mu^2} \tag{6}$$

where

l is the characteristic length of heat-transfer surface for horizontal plate, 0.035 (m)
ρ is the density of glycerol oil at reference temperature, 1225 (kg/m^3)
g is the gravitational acceleration (m/s^2)
β is the coefficient of thermal expansion of glycerol oil, 5.65×10^{-4} (1/K)
ΔT is the temperature difference of heated surface and ambient temperature, 70 (K)
μ is the dynamic viscosity of glycerol oil at reference temperature, 0.04 (kg/m·s)

The Prandtl number can be determined according to Equation (7) to be 377.

$$Pr = \frac{C_p \mu}{k_1} \tag{7}$$

where

C_p is the specific heat capacity of glycerol oil at reference temperature, 2680 (J/kg·K)
μ is the dynamic viscosity of glycerol oil at reference temperature, 0.04 (kg/m·s)
k_1 is the heat conductivity of glycerol oil at reference temperature, 0.284 (W/m·K)

Therefore, in the horizontal plate, the free convective heat-transfer coefficient of the glycerol oil bath, h_1, can be calculated to be 190 W/m^2·K.

The free convective heat-transfer coefficient for the zeolite gel in the non-stirred reactor, h_2, can be calculated according to Equations (8) and (9), when $Gr \cdot Pr$ is less than 2×10^8 [20].

$$Nu = 0.13(Gr \cdot Pr)^{1/3} \tag{8}$$

$$h_2 = \frac{k_2}{l} 0.13(Gr \cdot Pr)^{1/3} \tag{9}$$

where

k_2 is the heat conductivity of the zeolite gel at reference temperature (W/m·K)
l is the characteristic length of heat-transfer surface for horizontal plate (m)

The Grashof number and the Prandtl number in this case, can be calculated according to Equations (10) and (11), respectively.

$$Gr = \frac{l^3 \rho^2 g \beta \Delta T}{\mu^2} \tag{10}$$

$$Pr = \frac{C_p \mu}{k_2} \tag{11}$$

where

l is the characteristic length of the heat-transfer surface for horizontal plate, 0.035 (m)
ρ is the density of the zeolite gel at reference temperature, 1167 (kg/m³)
g is the gravitational acceleration (m/s²)
β is the coefficient of thermal expansion of the zeolite gel, 5.11×10^{-4} (1/K)
ΔT is the temperature difference of the heated surface and ambient temperature, 70 (K)
μ is the dynamic viscosity of the zeolite gel at reference temperature, 0.0023 (kg/m·s)
C_p is the specific heat capacity of the zeolite gel at reference temperature, 4190 (J/kg·K)
k_2 is the heat conductivity of the zeolite gel at reference temperature, 0.669 (W/m·K)

Thus, Gr and Pr are calculated as 3.87×10^6 and 14.4, respectively. Substituting k_2, l, Gr, and Pr, h_2 is calculated to be 949 W/m²·K. Therefore, in horizontal plate, the overall heat-transfer coefficient in the glycerol oil bath heating, $U_{oil\ bath}$, is calculated to be 154 W/m²·K.

The $U_{oil\ bath}$ at vertical plane/cylinder (side wall of the reactor) is determined based on Equation (2). The h_1 is calculated according to the relation of dimensionless group of Nu in Equations (12) and (13) when $10^4 \leq Gr \cdot Pr \leq 10^9$ ($Gr \cdot Pr = 2.37 \times 10^8$) [20].

$$Nu = 0.59(Gr \cdot Pr)^{1/4} \tag{12}$$

$$h_1 = \frac{k_1}{l} 0.59 (Gr \cdot Pr)^{1/4} \tag{13}$$

where

h_1 is the free convective heat-transfer coefficient of the glycerol oil for vertical plane/cylinder (Region 0–1) (W/m²·K)
k_1 is the heat conductivity of glycerol oil at reference temperature (W/m·K)
l is the characteristic length of heat-transfer surface for vertical plane/cylinder, 0.12 (m)

Thus, the h_1 is calculated to be 173 W/m²·K. The h_2 is calculated according to the relation of dimensionless group of Nu in Equations (14) and (15) for a vertical plane/cylinder when $10^9 \leq Gr \cdot Pr \leq 10^{13}$ ($Gr \cdot Pr = 2.25 \times 10^9$) [20].

$$Nu = 0.021(Gr \cdot Pr)^{2/5} \tag{14}$$

$$h_2 = \frac{k_2}{l} 0.021(Gr \cdot Pr)^{2/5} \tag{15}$$

where

h_2 is the free convective heat-transfer coefficient of the zeolite for vertical plane/cylinder (Region 0–1) (W/m²·K)
k_2 is the heat conductivity of zeolite at reference temperature (W/m·K)
l is the characteristic length of heat-transfer surface for vertical plane/cylinder, 0.12 (m).

Thus, the h_2 is calculated to be 645 W/m²·K. From Equation (2), the $U_{oil\ bath}$ in vertical plane/cylinder is 133 W/m²·K. Therefore, $U_{oil\ bath}$ of the total system can be determined by the proportion of the contact surface, is 138 W/m²·K.

When employing the magnetic induction heating system, there are two resistances to heat transfer as shown in Figure 6b: Region 1–2, in which heat is conducted through the stainless steel, and the heat transfer is governed by Fourier's law; Region 2–3, where the reactor bottom is in contact with the zeolite gel at ambient temperature ($T_{zeolite\ gel}$), and the heat transfer at the boundary is governed by Newton's law of cooling. Contrarily, in Region 0–1, the heat is generated by induction, and the reactor bottom is in contact with the bottom of the ferromagnetic vessel at constant temperature ($T_{induction}$). There is no resistance at this region.

The heat-transfer equation for the magnetic induction heating is given in Equation (16):

$$Q_{induction} = \frac{\Delta T_{induction} U_{induction}}{\frac{1}{A}} = \frac{\Delta T_{induction}}{\frac{L}{kA} + \frac{1}{h_3 A}} \tag{16}$$

where

$Q_{induction}$ is the amount of heat flow in the magnetic induction heating system (W)
$\Delta T_{induction}$ is the temperature difference between $T_{induction}$ and $T_{zeolite\ gel}$, 223 (K)
$U_{induction}$ is the overall heat-transfer coefficient in the magnetic induction heating system (W/m$^2 \cdot$K)
A is the area of the reactor bottom (m^2)
L is the thickness of the reactor bottom, 0.003 (m)
h_3 is the free convective heat-transfer coefficient of the zeolite gel for the non-stirred reactor (Region 2–3) (W/m$^2 \cdot$K)
k is the thermal conductivity of the stainless steel (Region 1–2), 16 (W/m·K)

$U_{induction}$ is calculated according to Equation (17):

$$U_{induction} = \frac{1}{\frac{L}{k} + \frac{1}{h_3}} \tag{17}$$

h_3 can be calculated according to Equations (18) and (19), when $Gr \cdot Pr < 2 \times 10^8$.

$$Nu = 0.13(Gr \cdot Pr)^{1/3} \tag{18}$$

$$h_3 = \frac{k_3}{l} 0.13(Gr \cdot Pr)^{1/3} \tag{19}$$

With the heat input of the induction is 600 W, heating time is 3 min, $T_{induction}$ is 521 K, and $T_{zeolite\ gel}$ is 298 K, $\Delta T_{induction}$ is calculated as 223 K. Substituting Gr, Pr, k_3, and l, h_3 is determined as 1396 W/m$^2 \cdot$K. Thus, the overall heat-transfer coefficient in the magnetic induction heating, $U_{induction}$, is calculated as 1106 W/m$^2 \cdot$K.

Finally, by substituting all values into Equation (16), the $Q_{induction}$ is calculated to be 3848 W, and $Q_{oil\ bath}$ is 664 W. The calculation shows that $Q_{induction}$ is greater than $Q_{oil\ bath}$ due to two factors: (1) the overall heat transfer coefficient and (2) the temperature difference between the heating source and zeolite gel. Firstly, $U_{induction}$ is eight times greater than $U_{oil\ bath}$ as the magnetic induction system does not depend on the free convective heat transfer of the glycerol oil heating medium. The overall heat-transfer coefficient is in analogy with the electrical conductance in the electrical circuit. The higher the electrical conductance, the higher the electrical current flow. Similarly, the higher the overall heat-transfer coefficient, the higher the heat transfer rate. Secondly, $\Delta T_{induction} > \Delta T_{oil\ bath}$ as the magnetic inductor can produce a surface temperature of the ferromagnetic vessel much greater than the oil temperature. The temperature difference in heat transfer is in analogy with the voltage potential in the electrical circuit. The higher the voltage, the higher the electrical current flow. Similarly, the higher the temperature difference, the higher the heat-transfer rate.

In addition, the energy supplies of both heating methods are determined based on one gram of zeolite production as 84.6 kJ for magnetic induction heating and 258 kJ for convection oil bath heating. It shows the advantage in energy saving by applying the induction heating comparing to the convection glycerol oil bath heating. The energy supply is three times lower.

As a result, the high heat-transfer rate produced by the magnetic induction heating could give an opportunity for the non-stirred method (without agitation) in the zeolite synthesis system especially for large-scale production. Typically, the large-scale synthesis needs stirring the zeolite gel, e.g., by impeller, during the reaction to increase the efficiency of the heat transfer in the reaction content. Therefore, the magnetic induction heating technique can create possibility to the successful synthesis of broad ranges of zeolites including the structural types that prefer to form in the static condition.

4. Conclusions

The magnetic induction assisted technique is an alternative heating method for the hydrothermal synthesis of zeolites with a high heat-transfer rate and a more uniform temperature distribution. This technique, which involves direct contact between the surface of the heat source and the reactor, exhibits a higher heat-transfer rate than that of the conventional convection oil bath technique. This is due to (1) the eight-time higher overall heat transfer coefficient, attributed to the absence of the resistance of the heating medium layer and (2) the higher temperature difference between the heating source and the zeolite gel. It results more energy saving (three-time lower) by the magnetic induction heating than that by the convection oil bath. The magnetic induction heating by direct surface contact shows promising application in the large-scale zeolite synthesis due to its efficient heat transfer and energy saving. The efficient heat transfer from the heating source to the zeolite gel provides more flexibility to the zeolite synthesis under non-stirred condition. Thereby, it can give possibilities for the successful synthesis of various zeolites in the large-scale production.

Author Contributions: S.T.—conceptualization, formal analysis, funding acquisition, investigation, methodology, visualization, writing-original draft preparation, writing-review and editing; C.P.—conceptualization, formal analysis, investigation, methodology, visualization, writing-original draft preparation; W.S.-n.—investigation, visualization; S.S.—investigation, visualization. All authors have read and agreed to the published version of the manuscript.

Funding: This research was funded by King Mongkut's University of Technology North Bangkok, grant number KMUTNB-ART-60-053.

Institutional Review Board Statement: Not applicable.

Informed Consent Statement: Not applicable.

Data Availability Statement: Data sharing is not applicable to this article.

Conflicts of Interest: The authors declare no conflict of interest.

References

1. Breck, D.W. *Zeolite Molecular Sieves: Structure, Chemistry, and Use*; John Wiley & Sons: New York, NY, USA, 1974; pp. 1–28.
2. Maesen, T. The zeolite scene—An overview. In *Studies in Surface Science and Catalysis*; Čejka, J., Van Bekkum, H., Corma, A., Schüth, F., Eds.; Elsevier: Amsterdam, The Netherlands, 2007; Volume 168, pp. 1–12.
3. Osinga, T.J.; Dekker, J.N.P.M. Zeolites in Liquid Detergent. Compositions. Patent WO89/04360, 18 May 1989.
4. Dee, D.P.; Chiang, R.L.; Miller, E.J.; Whitley, R.D. High Purity Oxygen Production by Pressure Swing Adsorption. U.S. Patent US6544318 B2, 8 April 2003.
5. Litz, J.E.; Williams, C.S. Radium Removal from Aqueous Media Using Zeolite Materials. U.S. Patent US9908788 B1, 6 March 2018.
6. Verduijn, J.P.; Mohr, G. Zeolite Catalyst and Its Use in Hydrocarbon Conversion. U.S. Patent WO97/45198, 4 December 1997.
7. Kuehl, G.H.; Rosinski, E.J. Catalytic Cracking with Framework Aluminium. U.S. Patent US4954243, 4 September 1990.
8. Li, Y.; Li, L.; Yu, J. Applications of zeolites in sustainable chemistry. *Chem* **2017**, *3*, 928–949. [CrossRef]
9. Elangovan, S.P.; Ogura, M.; Ernst, S.; Hartmann, M.; Tontisirin, S.; Davis, M.E.; Okubo, T. A comparative study of zeolites SSZ-33 and MCM-68 for hydrocarbon trap applications. *Microporous Mesoporous Mater.* **2006**, *96*, 210–215. [CrossRef]
10. Tontisirin, S. Highly crystalline LSX zeolite derived from biosilica for copper adsorption: The green synthesis for environmental treatment. *J. Porous Mater.* **2015**, *22*, 437–445. [CrossRef]
11. Kongnoo, A.; Tontisirin, S.; Worathanakul, P.; Phalakornkul, C. Surface characteristics and CO_2 adsorption capacities of acid-activated zeolite 13X prepared from palm oil mill fly ash. *Fuel* **2017**, *193*, 385–394. [CrossRef]

12. Moliner, M.; Corma, A. From metal-supported oxides to well-defined metal site zeolites: The next generation of passive NO_x adsorbers for low-temperature control of emissions from diesel engines. *React. Chem. Eng.* **2019**, *4*, 223–234. [CrossRef]
13. Cundy, C.S.; Cox, P.A. The hydrothermal synthesis of zeolites: History and development from the earliest days to the present time. *Chem. Rev.* **2003**, *103*, 663–701. [CrossRef] [PubMed]
14. Cundy, C.S.; Cox, P.A. The hydrothermal synthesis of zeolites: Precursors intermediates and reaction mechanism. *Microporous Mesoporous Mater.* **2005**, *82*, 1–78. [CrossRef]
15. Yu, J. Synthesis of zeolite. In *Studies in Surface Science and Catalysis*; Cejka, J., Van Bekkum, H., Corma, A., Schüth, F., Eds.; Elsevier: Amsterdam, The Netherlands, 2007; Volume 168, pp. 39–102.
16. Lucía, O.; Maussion, P.; Dede, E.J.; Burdio, J.M. Induction heating technology and its applications: Past developments, current technology, and future challenges. *IEEE Trans. Ind. Electron.* **2014**, *61*, 2509–2520. [CrossRef]
17. Slangen, P.M.; Jansen, J.C.; Van Bekkum, H. Induction heating: A novel tool for zeolite synthesis. *Zeolites* **1997**, *18*, 63–66.
18. Maraş, T.; Nerat, E.Y.; Erdem, A.; Tatlier, M. Preparation of zeolite coating by induction heating of the substrate. *J. Sol.-Gel. Sci. Technol.* **2021**, *98*, 54–67. [CrossRef]
19. Treacy, M.M.J.; Higgins, J.B. *Collection of Simulated XRD Powder Patterns for Zeolites*, 5th ed.; Elsevier: Amsterdam, The Netherlands, 2007; pp. 170–171.
20. Holman, J.P. *Heat Transfer*, 6th ed.; McGraw-Hill: New York, NY, USA, 1986; pp. 323–371.
21. McCabe, W.L.; Smith, J.C.; Harriott, P. *Unit Operations of Chemical Engineering*, 7th ed.; McGraw-Hill: Boston, MA, USA, 2005; pp. 1093–1107.
22. Glycerine Producers' Association. *Physical Properties of Glycerine and Its Solutions*; Glycerine Producers' Association: New York, NY, USA, 1963; pp. 1–27.
23. Poling, B.E.; Thomson, G.H.; Friend, D.G.; Friend, D.G.; Rowley, R.L.; Wilding, W.V. Physical and chemical data. In *Perry's Chemical Engineers' Handbook*, 8th ed.; Green, D.W., Perry, R.H., Eds.; McGraw-Hill: New York, NY, USA, 2008; pp. 2-1–2-517.

Article

The Influence of Loop Heat Pipe Evaporator Porous Structure Parameters and Charge on Its Effectiveness for Ethanol and Water as Working Fluids

Krzysztof Blauciak, Pawel Szymanski * and Dariusz Mikielewicz

Faculty of Mechanical Engineering and Shipbuilding, Gdansk University of Technology, Narutowicza 11/12, 80-233 Gdansk, Poland; k.blauciak@frigoconsulting.com (K.B.); dmikiele@pg.edu.pl (D.M.)
* Correspondence: pawszym1@pg.edu.pl

Abstract: This paper presents the results of experiments carried out on a specially designed experimental rig designed for the study of capillary pressure generated in the Loop Heat Pipe (LHP) evaporator. The commercially available porous structure made of sintered stainless steel constitutes the wick. Three different geometries of the porous wicks were tested, featuring the pore radius of 1, 3 and 7 µm. Ethanol and water as two different working fluids were tested at three different evaporator temperatures and three different installation charges. The paper firstly presents distributions of generated pressure in the LHP, indicating that the capillary pressure difference is generated in the porous structure. When installing with a wick that has a pore size of 1 µm and water as a working fluid, the pressure difference can reach up to 2.5 kPa at the installation charge of 65 mL. When installing with a wick that has a pore size of 1 µm and ethanol as a working fluid, the pressure difference can reach up to 2.1 kPa at the installation charge of 65 mL. The integral characteristics of the LHP were developed, namely, the mass flow rate vs. applied heat flux for both fluids. The results show that water offers larger pressure differences for developing the capillary pressure effect in the installation in comparison to ethanol. Additionally, this research presents the feasibility of manufacturing inexpensive LHPs with filter medium as a wick material and its influence on the LHP's thermal performance.

Keywords: Loop Heat Pipe; porous materials; mass transfer; heat transfer; phase transitions

Citation: Blauciak, K.; Szymanski, P.; Mikielewicz, D. The Influence of Loop Heat Pipe Evaporator Porous Structure Parameters and Charge on Its Effectiveness for Ethanol and Water as Working Fluids. *Materials* **2021**, *14*, 7029. https://doi.org/10.3390/ma14227029

Academic Editor: Anatoliy Pavlenko

Received: 26 October 2021
Accepted: 17 November 2021
Published: 19 November 2021

Publisher's Note: MDPI stays neutral with regard to jurisdictional claims in published maps and institutional affiliations.

Copyright: © 2021 by the authors. Licensee MDPI, Basel, Switzerland. This article is an open access article distributed under the terms and conditions of the Creative Commons Attribution (CC BY) license (https://creativecommons.org/licenses/by/4.0/).

1. Introduction

LHPs are very efficient heat transfer devices operating passively where the principle of operation is based on evaporation and condensation of the working fluid at a specific pressure related to the required conditions. In such a two-phase passive thermal control apparatus, extensive amounts of heat can be transferred with stable control of the heat source temperature. There has been a widespread effort to extend successful applications of LHPs to more common terrestrial applications [1–8] in order to develop more passive cooling systems, mainly based on liquid–vapour phase-change mechanisms to remove large heat fluxes. The electronic terrestrial applications benefit from the cooling advantages of LHPs (e.g., passive—electrical power-free, long-distance heat transfer, flexibility in design and assembly, robustness, antigravity capability, noise and vibration-free operation).

The demand for cooling and thermal management of electronic devices increases over the limits of the current state-of-the-art cooling technologies, which primarily result in challenges towards miniaturisation of electronics and transfer higher heat fluxes from the electronic components produced at the present day by the space and terrestrial electronics industry.

The wick structure installed in the evaporator is responsible for providing a high capillary pressure to circulate the working fluid in the system. The most important parameters that characterise the wicks are permeability, thermal conductivity, capillary pumping performance, effective pore radius, interface heat transfer and wettability [9–11]. These parameters are determined by the internal wick structure and material properties and depend on the manufacturing process of the wick itself. According to the literature [9,12], most porous structures used in LHPs have been made of metallic materials, such as nickel, titanium, aluminium, stainless steel and, occasionally, ceramic, polymer and either silicone or foam. The most widespread technology for the manufacturing of metal wicks is sintering.

Nowadays, several laboratories endeavour to find a novel method of fabrication of wick or new materials which provide high capillary forces and high permeability or mass flow rate (e.g., additive manufacturing (AM)—colloquially known as 3D printing) [9–11,13–15], as these two design features are typically inhibitive of each other. AM is a very promising method of wick or LHP manufacturing; however, it is still costly and needs a lot of research to be conducted in this area. The main drawback is currently the minimum pore size that can be manufactured using AM, which limits the use of AM LHPs in long-distance transport applications.

Some scholars make attempts to manufacture a low-cost, functional LHP [16] or utilise the porous material manufactured by filter appliances companies as an LHP wick (e.g., Siedel [17]). Some attempt to use commercially available porous structures such as, for example, stainless steel sintered porous structures. Hence, this research is a continuation of the work carried on by Mikielewicz et al. [18–20] and presents the feasibility of manufacturing economical LHPs with sintered stainless steel powder by Tridelta Siperm GmbH (Dortmund, Germany) [21] as a wick material and its influence on LHPs thermal performance.

This paper presents studies of the capillary effect in commercially available stainless steel porous structures with different pore sizes of which the wick of the LHP evaporator is made. Two working fluids were considered in the tests, namely water and ethanol, at three different evaporation temperatures. Different installation charges were considered and compared to find an appropriate amount of working fluid inventory and its influence on LHP thermal performance. Characteristics of the distributions of pressure increase and mass flow rate in the function of heat flux were presented and discussed.

2. Experimental Rig

The LHP evaporator with the sintered stainless steel porous wick was manufactured, assembled and tested to evaluate the possible capillary pressure difference created by the porous structure within the evaporator and its thermal performance. The evaporator was designed to enable the wick to be exchanged to a different one with another pore size. The experiment consisted of measuring the pressure rise in the evaporator while changing the applied heat load to the evaporator casing, resulting in different thermal and operational conditions. The test facility is schematically presented in Figure 1.

The principle of LHP operation is rather straightforward [1,2,20]. When heat is supplied to the evaporator, the meniscus is formed at the liquid/vapor interface in the evaporator wick, generating the required capillary forces to pump the fluid. Surface tension developed in a wick is a source of the pumping force used to circulate the fluid in the loop. The produced vapor flows down through the system of grooves then to the evaporator, where the capillary pressure pushes out the vapor in the direction of the vapor line towards the condenser rendering the fluid transport around the loop. The compensation chamber (CC) serves for storing and sustaining the surplus of working fluid and control of LHP operation.

Considering the large variety of working fluids possible to apply in the LHP installation, it was decided to use the most common and previously applied working fluids in LHPs. Based on this analysis, the rig design requirements, physical properties of the working fluid and its possibility of generating the largest capillary pressures for further experiments were selected two fluids, namely distilled water and technical-grade ethanol. Such fluids are suitable for analysis due to their high potential to generate a capillary pressure difference in the porous structure (Δp_c). A capillary pressure difference ($\Delta p_c = 2\sigma/R_p$)

depends on the surface tension of the working fluid (σ) and pore radius (R_p). Characteristics of the theoretically feasible capillary pressure rise as a function of temperature for water are presented in Figure 2 and for ethanol in Figure 3. Both cases presented the effect of pore size on the generated pressure rise. In the considered temperature range, it can be seen that water has a higher potential to create a capillary pressure rise for each of the considered wick pore radiuses. The highest values of Δp_c are obtained for the smallest values of the pore radius. With increasing temperature, the potential to generate Δp_c decreases.

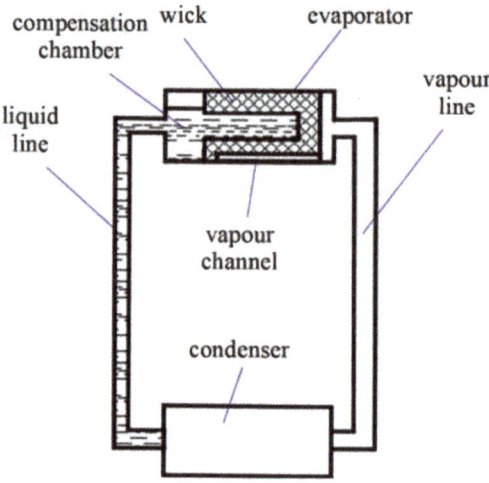

Figure 1. The layout of LHP [20].

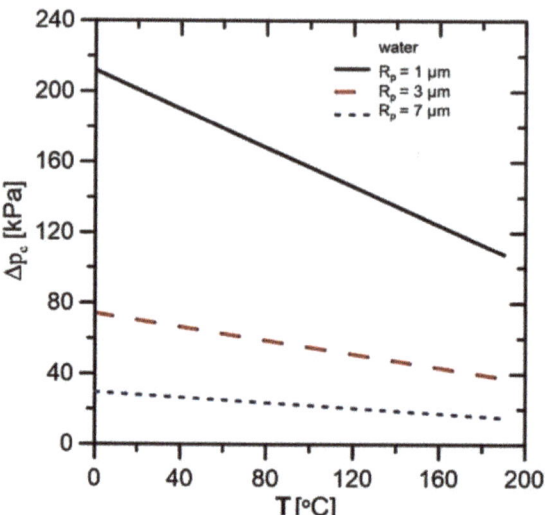

Figure 2. Capillary pressure rise characteristic for water as a working fluid.

Table 1 presents the basic physical properties of the selected working fluids. The data were determined using the REFPOROP 10.0 software [22]. In the installation filled with water, triple distilled water was used to avoid undesirable corrosion effects or the formation of sediments inside the LHP elements during the evaporation and condensation processes, while in the case of the second working fluid, technical-grade ethanol with a concentration of 99.7% was used.

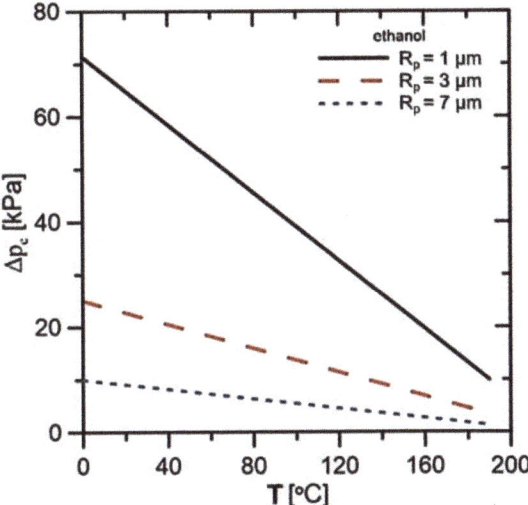

Figure 3. Capillary pressure rise characteristic for ethanol as a working fluid.

Table 1. Comparison of the basic physical properties of water (H_2O) and ethanol (C_2H_5OH).

Working Fluid	Critical Pressure	Critical Temperature	Molar Mass	Triple Point Temperature	Boiling Temperature
	[kPa]	[°C]	[kg/kmol]	[°C]	[°C]
H_2O	22.064	373.95	18.015	0.01	99.974
C_2H_5OH	6268	241.56	46.068	−114.15	78.420

In order to illustrate the changes in individual physicochemical parameters and dimensional numbers as a function of temperature, an extract of the values of selected physical properties of fluids for several selected operating temperatures of the LHP is presented (Table 2). The data were determined using the REFPROP 10.0 software.

Table 2. Comparison of physicochemical parameters of water and ethanol versus temperature.

Working Fluid	Liquid Density	Vapor Density	Specific Heat of Liquid	Specific Heat of Vapor	Enthalpy of Vaporisation	Liquid Viscosity	Vapor Viscosity	Prandtl Number of Liquid	Prandtl Number of Vapor	Surface Tension
	[kg/m³]	[kg/m³]	[kJ/kgK]	[kJ/kgK]	[kJ/kg]	[μPas]	[μPas]			[N/m]
$H_2O\mid_{T=20}$	998.16	0.017314	4.1844	1.9059	2453.5	1001.6	9.5441	7.0038	0.9979	0.072736
$H_2O\mid_{T=40}$	992.18	0.051242	4.1796	1.9314	2406.0	652.72	10.185	4.3263	1.0037	0.069596
$H_2O\mid_{T=80}$	971.77	0.293670	4.1969	2.0120	2308.0	354.04	11.539	2.2177	1.0089	0.062673
$H_2O\mid_{T=100}$	958.35	0.598170	4.2157	2.0800	2256.4	281.58	12.232	1.7480	1.0138	0.058912
$H_2O\mid_{T=120}$	943.11	1.122100	4.2435	2.1770	2202.1	232.03	12.927	1.4412	1.0245	0.054968
$C_2H_5OH\mid_{T=20}$	789.59	0.112410	2.5121	1.5840	926.61	1195.2	8.6186	18.054	0.7826	0.022414
$C_2H_5OH\mid_{T=40}$	772.47	0.321860	2.7565	1.6488	904.94	821.65	9.2312	14.009	0.8082	0.019886
$C_2H_5OH\mid_{T=80}$	734.64	1.759100	3.2036	1.8150	846.97	429.47	10.431	9.0044	0.8542	0.015030
$C_2H_5OH\mid_{T=100}$	713.14	3.530000	3.4048	1.9319	809.83	322.63	110.22	7.4024	0.8785	0.012713
$C_2H_5OH\mid_{T=120}$	689.39	6.568700	3.5983	2.0846	766.47	246.81	116.14	6.1681	0.9052	0.010481

In this study, the evaporator filled with three different wick materials was tested. The evaporator design allows for the exchange of the porous wick. The outline of the evaporator is presented in Figure 4. A sintered stainless steel cylindrical wick material was manufactured by Tridelta Siperm GmbH, a provider of porous metals, and inserted inside the evaporator casing (Figure 5) [20]. The porous wicks have a mean pore radius of 1 µm, 3 µm and 7 µm, and porosity of 24%, 33% and 35%, respectively. Such material was selected for its high resistance to corrosion and chemical compatibility with water and ethanol [23]. The evaporator casing was made of copper. On the internal side of the evaporator's casing, 12 longitudinal vapor grooves necessary for transporting vapor to the evaporator outlet were drilled. The cross-section of the evaporator casing is presented in the photo (Figure 6) and the schematic (Figure 7). The entire length of the evaporator is 216.5 mm.

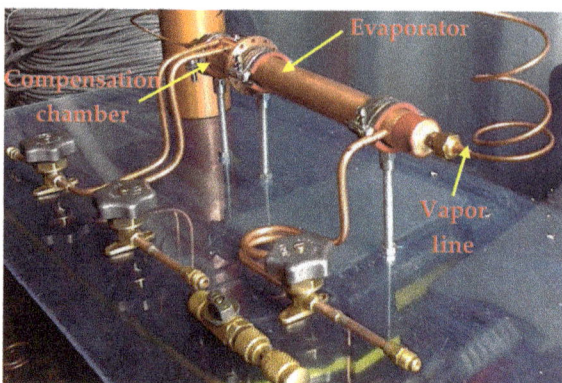

Figure 4. Photograph of the evaporator.

Figure 5. Examples of the wick manufactured by Tridelta Siperm GmbH [20].

One of the most difficult to design and the most important elements of LHP is the CC, as it is responsible for the control of pressure and temperature in the system as well as hydrodynamics within the loop. Following several tests, the volume of the CC was set to 0.043 dm^3 for the assumed dimensions of the evaporator. The sealed flange connection combined the evaporator and the CC, guaranteeing tightness and the possibility to exchange the wick and perspective evaporator revisions. Inside the CC, two thermocouples were installed to observe the temperature gradient inside during the application of LHP under different thermal loads.

Figure 6. Cross-section of evaporator casing [20].

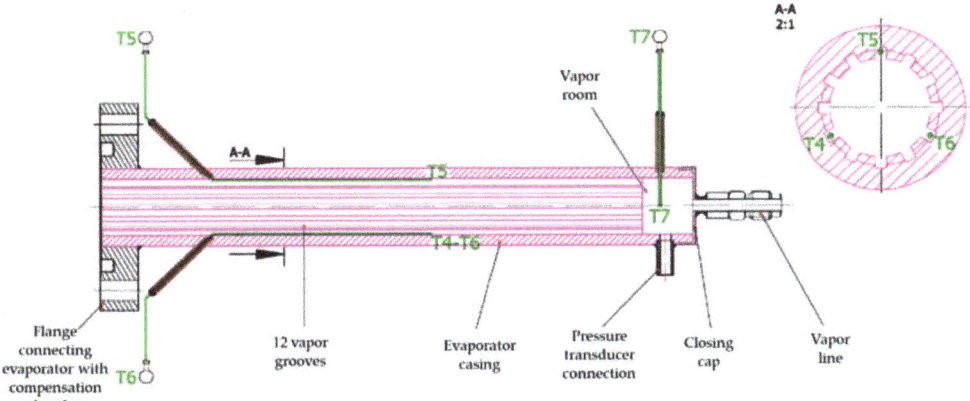

Figure 7. View of the evaporator with connections to thermocouples and pressure transducer.

The thermal load was applied using the electrical resistance wire. The wire was wound around the evaporator casing and connected to the laboratory DC supplier with adjustable current and voltage. Knowledge of the latter enabled calculation of the effective electric power applied to the resistor. Assuming that there was no heat loss through the insulation in the heating zone, the applied electrical power was taken as the rate of heat supplied to the system. Transport lines were made of smooth-wall copper tubes. The length of the liquid line length was 1152 mm (including bayonet) with an internal diameter equal to 2 mm. The vapor line length was 880 mm with an internal diameter of 2.95 mm. Condenser cooling was obtained using water circulating in a closed loop. Circulation of water was provided using a circulation pump featuring a flow rate up to 0.175 L/min. In parallel, the experimental rig was equipped with a visualisation section made from a transparent glass tube enabling the inspection of working fluid flow structures. Therefore, a two-way valve was installed to enable direct working fluid flow either through the copper vapor line or the transparent vapor line.

Figure 8 presents the outline of the LHP experimental facility where the location of all measuring points is indicated, whereas Figure 9 shows is the general view of the rig.

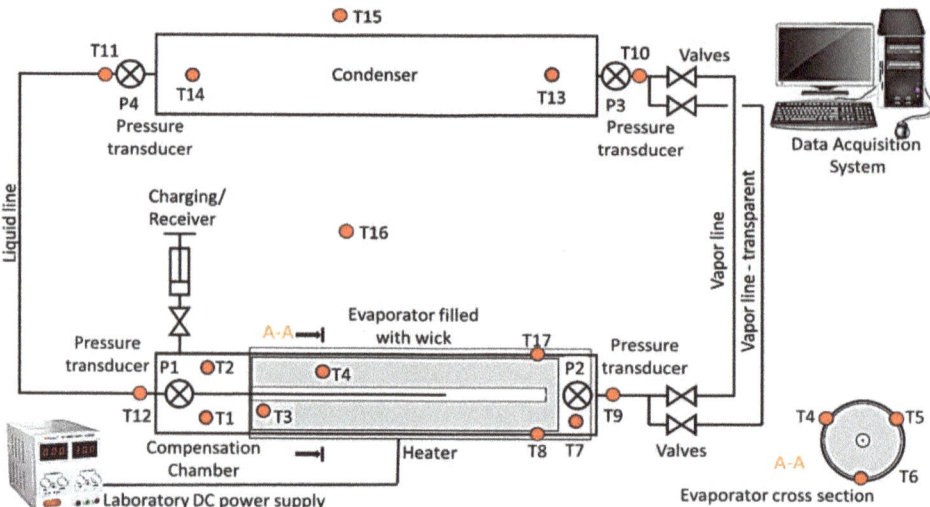

Figure 8. General schematic of LHP experimental rig with an indication of the measurement points.

Figure 9. Photograph of the test facility with marked measurement points (before insulation): 1—compensation chamber; 2—evaporator; 3—vapour line; 4—glass tube for inspection of flow structure in vapour line; 5—condenser; 6—liquid line; P1, P2, P3, P4—pressure transducers.

The following measurement instrumentation was performed using:

1. Eighteen class 1, T-type thermocouples with measurement error equal to $\pm(2.0 \times 10^{-3} \times [T] + 0.3\,°C + \text{number})$ integrated with multiplexer EMT200. Locations of thermocouples are shown in Figures 7 and 8;
2. Four class 0.1 pressure transducers with the measurement range of 1000 kPa of absolute pressure. Locations of pressure transducer measurement points are presented in Figures 8 and 9;
3. Digital multimeter UT71E for the recording of electrical power supplied to the evaporator through the resistance wire (measurement accuracy $\pm(2\% + 50)$).

Before measurements, the test rig was insulated using mineral wool with an aluminium coating to reduce the heat dissipation from the evaporator test section. In addition, the liquid and vapor lines were insulated using synthetic rubber of 13 mm thickness.

Each of the measurements was proceeded by installation vacuuming using a two-stage vacuum pump (model CPS VP6D). The maximum possible vacuum level of 99.999% (1.95×10^{-3} kPa abs) was normally achieved. Before the LHP startup, the condenser cooling section was filled with water. All measuring devices were connected and linked to the camera recorder. Then, using a special applicator, the installation was charged with working fluid. In order to obtain reproducible condensation conditions, the condenser chiller was initiated before the heating section until the desired steady temperature was obtained. The next steps consisted of a setup of the heating section recorder, launching the automatic temperature transducers, time and heating section recorders (with an automatic measurement recorder at 1 s intervals) and a recorder of pressure values displayed on panel displays (with a camera recording at 60 s intervals). The average measurement duration was determined to be about 120–130 min. Each of the measurement series was carried out for three evaporator temperature settings with the same working fluid volume. Therefore, each subsequent measurement required the use of cooling water in the thermostat with a similar temperature level. After the measurement with the last third heater setting, the installation was emptied from the working fluid, and then the above-mentioned procedure was repeated from the beginning by changing the parameters according to the previously described configuration of the test procedure.

The measurement uncertainties were estimated based on the analysis of systematic component errors of the measurement system [24] and presented in Table 3.

Table 3. List of measurement errors.

	Error Designation
Temperature	$\pm 0.2\,°C$
Pressure	± 25 Pa
Heat input	± 1.3 W
Working fluid volume	± 1 mL
Mass flow rate	$\pm 2 \times 10^{-5}$ kg/s

3. Results

As mentioned earlier, the experimental analysis consisted of experiments at three different working fluid charges, 60 mL, 65 mL and 70 mL, three different wicks featuring pores of 1 μm, 3 μm and 7 μm, and three different evaporator casing temperatures, $T_w = 90\,°C$, $T_w = 100\,°C$ and $T_w = 110\,°C$. Two different working fluids were tested: water and ethanol. In total, 12 saturated pressure levels were recorded in the case of water and 18 saturation pressure levels in the case of ethanol, while the thermal load was varied. This paper presents the results of pressure difference possible to reach in the evaporator with a porous wick for two installation charging ratios and two different fluids. The range of investigated parameters is presented in Table 4. The subsequent discussion is given for the case of a single filling volume of 65 mL and two test fluids.

Table 4. Range of investigated parameters.

Working Fluid	Charge Level [mL]	Heater Temp [°C]	Pore Size [μm]	P1–P2 [kPa]
WATER	65	90	1	2.5
		90	3	1.0
		100	1	2.2
		100	3	1.4
		110	1	0
		110	3	1.7
	70	90	1	1.5
		90	3	0.7
		100	1	1.6
		100	3	1.0
		110	1	0.9
		110	3	0
ETHANOL	65	90	1	1.2
		90	3	1.2
		90	7	1.5
		100	1	2.1
		100	3	2.0
		100	7	2.0
		110	1	1.9
		110	3	1.7
		110	7	0.3
	70	90	1	0.7
		90	3	1.1
		90	7	0.2
		100	1	0.8
		100	3	0.9
		100	7	0.2
		110	1	1.0
		110	3	0.8
		110	7	0.4

In the case of water, the results of pressure distributions are presented in Figures 10–13 for the charge volume of 65 mL. Analysis of the developed pressure in the case of water as a working fluid shows the pressure values before and after the evaporator, which vary with respect to the applied evaporator saturation temperature and the pore size. The level of pressure drop in the vapor and liquid lines is much smaller with respect to the pressure difference P2–P1, where P2 is the pressure after the evaporator (the highest pressure in the loop). P1 is the pressure before CC (the lowest pressure in the loop). Some pressure fluctuations are observed in all pressure distributions; however, consistent pressure differences are noticed in the distributions. During the investigations, the measurement run lasted for about 2.5 h, from which we can detect that reaching the steady-state conditions lasted for the first 30 min.

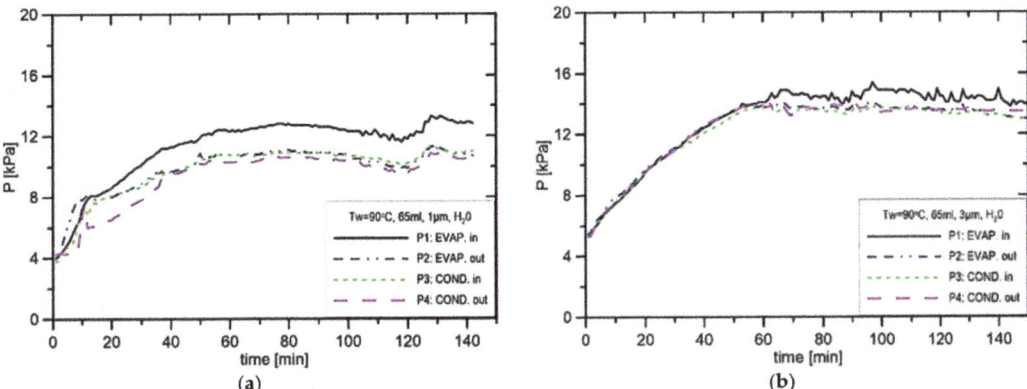

Figure 10. Distribution of pressure in function of time in case of water, charge volume 65 mL, $T_w = 90\ °C$, (**a**) $R_p = 1\ \mu m$, (**b**) $R_p = 3\ \mu m$.

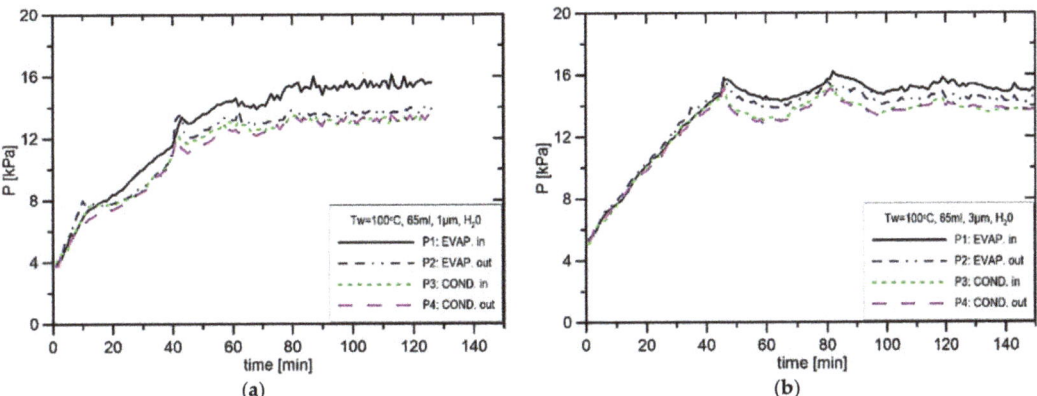

Figure 11. Distribution of pressure in function of time in case of water, charge volume 65 mL, $T_w = 100\ °C$, (**a**) $R_p = 1\ \mu m$, (**b**) $R_p = 3\ \mu m$.

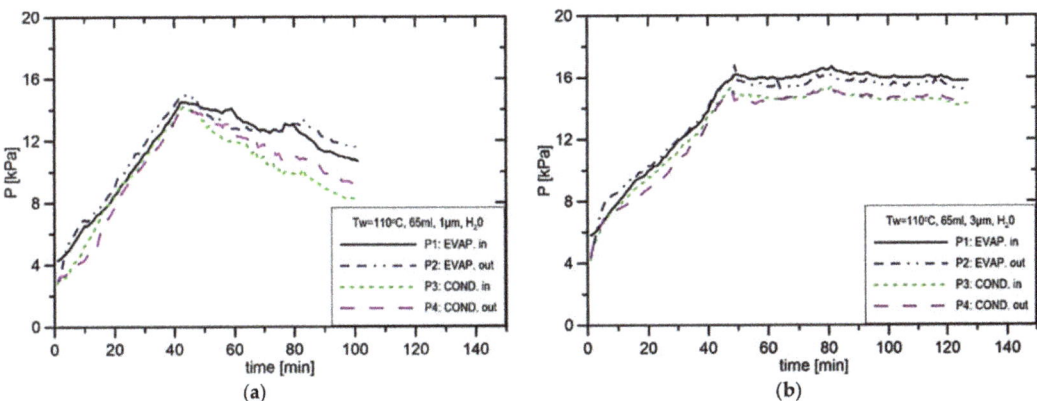

Figure 12. Distribution of pressure in function of time in case of water, charge volume 65 mL, $T_w = 110\ °C$, (**a**) $R_p = 1\ \mu m$, (**b**) $R_p = 3\ \mu m$.

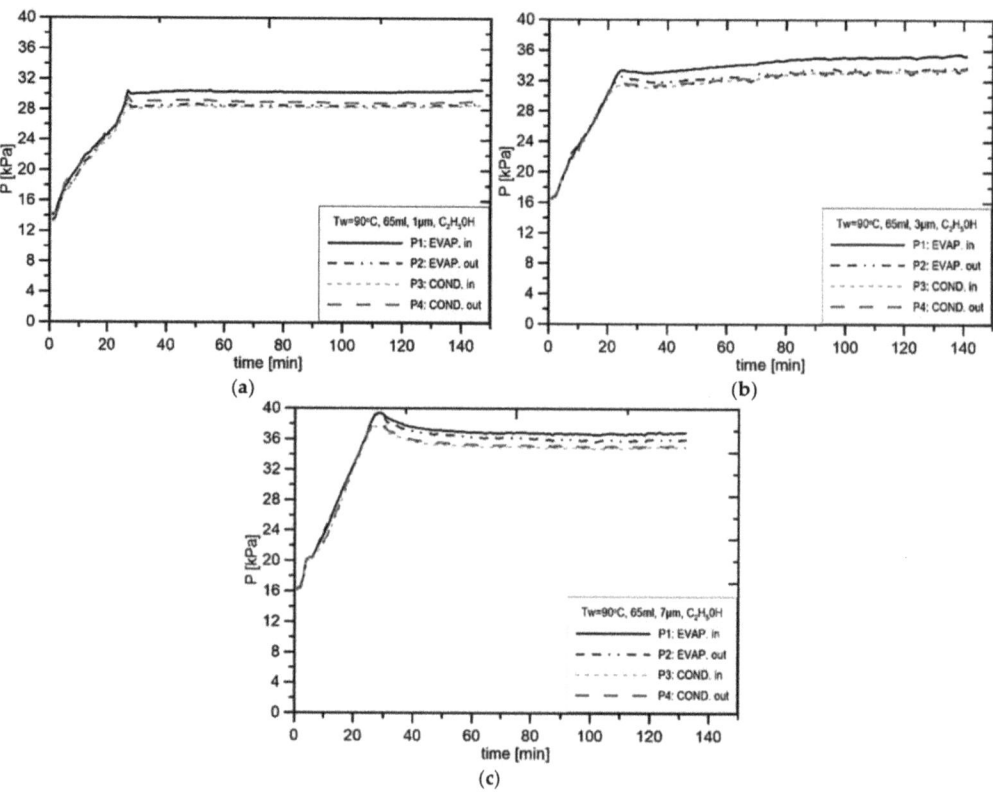

Figure 13. Distribution of pressure in function of time when using ethanol, charge volume 65 mL, $T_w = 90\ °C$, (**a**) R = 1 μm, (**b**) $R_p = 3$ μm, (**c**) $R_p = 7$ μm.

In Figure 10a, the rate of pressure increase is about 0.44 kPa/s, and the maximum pressure difference P1–P2 was equal to 2.55 kPa. In the experiment presented in Figure 10b, the pressure P1 stabilises at the level of 14.1 kPa, while in the vapor line, it stabilises at 12.9 kPa. The only difference between the experiment presented in Figure 10a,b is the size of the pore, which was changed from 1 μm to 3 μm. That confirms the fact that the reduction in the pore size leads to an increase in produced pressure. Figures 10a and 11a presented a similar situation to those presented in Figures 10b and 11b; however, the evaporator casing temperature increased from 90 °C to 100 °C. In this case, the pressure difference in the installation P1–P2 amounts merely to 2.2 kPa. In the experiment presented in Figure 11a, the pressure in the CC settled at the level of 15.0 kPa and 13.6 kPa in the vapor line, which results in the maximum pressure difference in the installation of 1.4 kPa. This suggests that the increase in evaporator casing temperature from 90 °C to 100 °C reduces the potential to produce the capillary pressure difference. In the experiments presented in Figure 12a,b, the evaporator casing temperature was set to 110 °C, whereas the pore size was equal to 1 μm and 3 μm, respectively. In the case of the run presented in Figure 12b, the pressure in the vapor line was 14.2 kPa, and the pressure in the CC was 15.9 kPa, which results in the maximum pressure difference in the installation equal to 1.7 kPa.

The analysis of the pressure distribution for ethanol as a working fluid and filling volume of 65 mL was presented in Figures 13–15. The general observation is that the pressure drops in the vapor and liquid lines are smaller in comparison to the pressure difference in the evaporator. Another observation is that the pressure fluctuations are generally smaller than in the case of water as a working fluid, although the operating pressure of the loop is much higher than in the case of water.

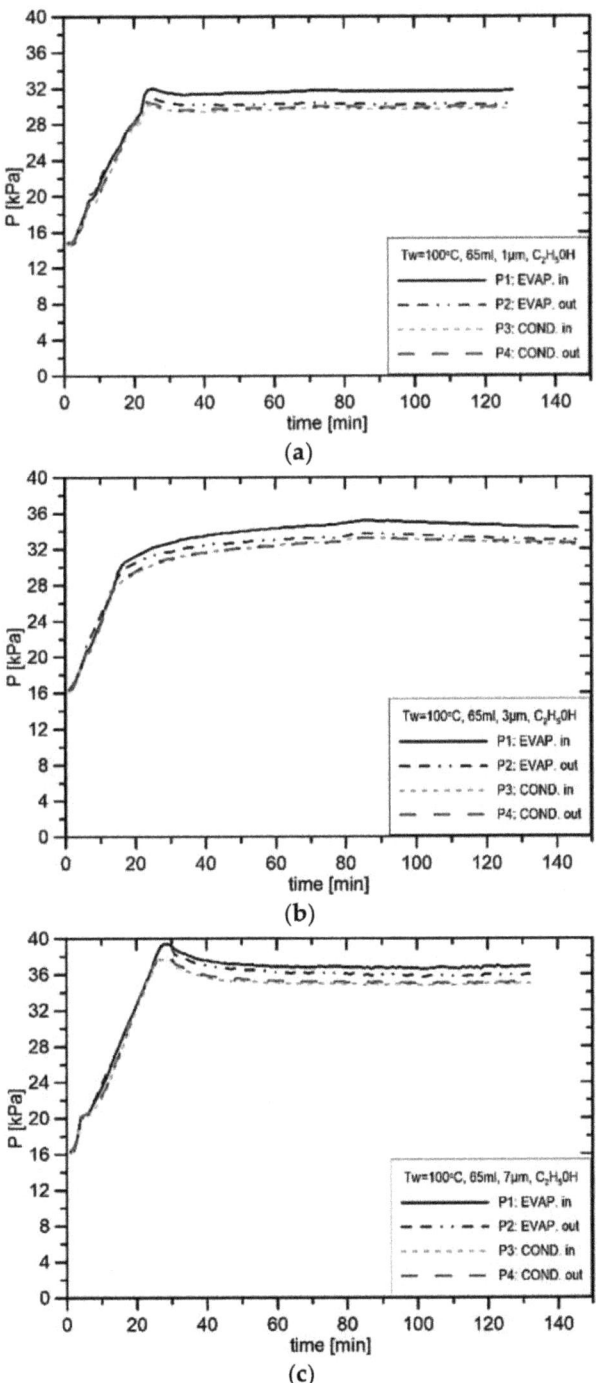

Figure 14. Distribution of pressure in function of time when using ethanol, charge volume 65 mL, $T_W = 100\ °C$, (**a**) $R_p = 1\ \mu m$, (**b**) $R_p = 3\ \mu m$, (**c**) $R_p = 7\ \mu m$.

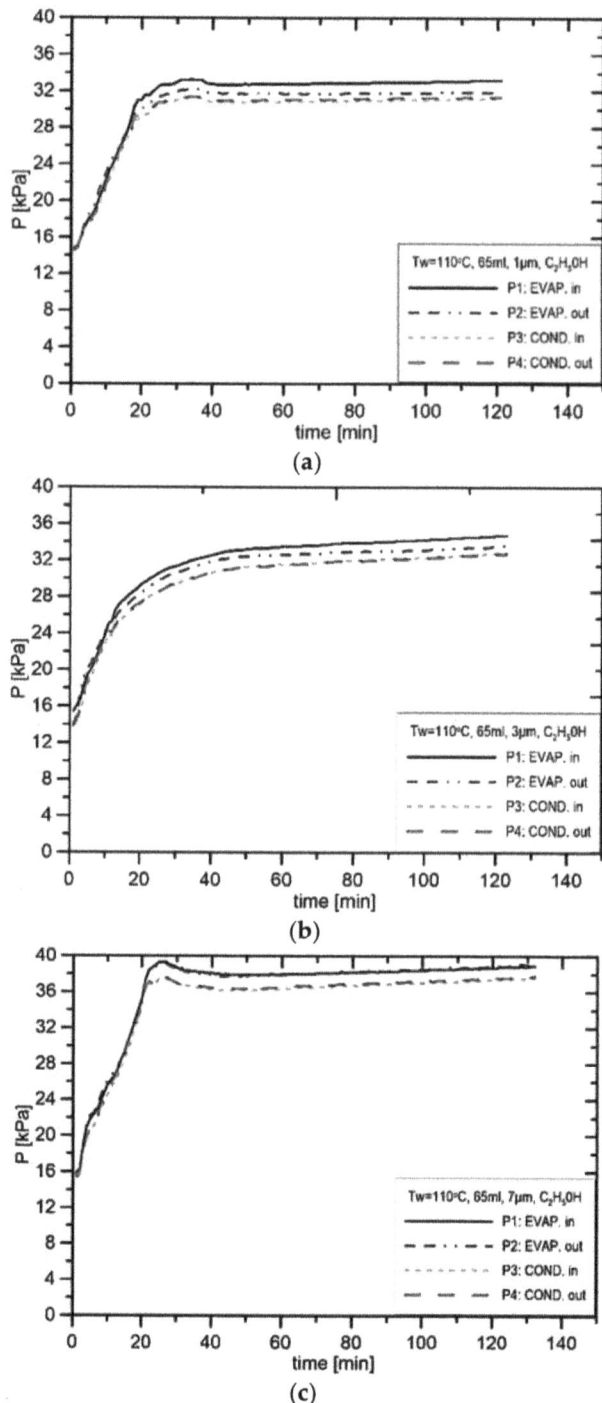

Figure 15. Distribution of pressure in function of time when using ethanol, charge volume 65 mL, $T_W = 110\ °C$, (**a**) $R_p = 1\ \mu m$, (**b**) $R_p = 3\ \mu m$, (**c**) $R_p = 7\ \mu m$.

For experiments with ethanol as a working fluid, the experimental run lasted for 2.5 h, and 30 min was related to the startup. The difference from the case of water was much higher initial pressure in the installation.

In Figure 13a, the initial pressure stabilised at the level of about 13.5 kPa, and due to the heat supply, the maximum pressure reached 30.2 kPa. In the case of other pore sizes as well as other evaporator casing temperatures, the pressure reached values of 40.0 kPa. In the case of the parameters presented in Figure 13b, the pressure in the CC settled at the level of 30.2 kPa, while in the vapor line, it settled at the level of 29.0 kPa. Therefore, the wick produces a pressure increase of 1.2 kPa for the evaporator casing temperature equal to 90 °C and the pore size of 1 µm. In the case of the experimental run presented in Figure 13b, where the pore size is equal to 3 µm, the pressure before the CC settled at 35.1 kPa, and the pressure in the vapor channel settled at the level of 33.9 kPa. This confirms that the reduction in the pore radius leads to an increase in pressure. Figure 13c presents the results for the case when the wick pore size is 7 µm. In that case, the pressure before the CC stabilised at the level of 37.1 kPa, whereas the pressure in the vapor line was at the level of 35.6 kPa. This indicates that the size of the pores has practically no influence on the capillary pressure difference.

Figure 14 presents the results of experiments where the evaporator casing temperature is 100 °C at the filling volume of 65 mL. From the results in Figure 14a, the pressure in the CC is 32.0 kPa, and the pressure in the vapor line is 29.9 kPa; hence, the pressure difference at the installation is 2.1 kPa. In the case presented in Figure 14b, the pressure in the CC settled at the level of 35.0 kPa, and in the vapor line, it settled at the level of 33.0 kPa, which gives a pressure difference in the installation of 2.0 kPa. This pressure difference is smaller than in the case of the pressure difference from Figure 14a, which confirms the fact that with the increase in the pore size, the potential to produce capillary pressure difference decreases. In the case of pore size equal to 7 µm (Figure 14c), the pressure in the CC stabilised at the value of 37.0 kPa, and in the vapor line, it stabilised at the level of 35.0 kPa. In Figure 15 presented are the results of experimental runs with the evaporator casing temperature set to 110 °C and three different pore sizes of 1 µm, 3 µm and 7 µm. A similar character of changes as in the case of evaporator temperature settings of 90 °C and 100 °C is present. From the comparison of the three values of heater setting, with the increase in evaporator casing temperature, at the same value of the pore size, the potential to produce the capillary temperature difference decreases. Such a conclusion can be drawn by comparing Figures 13a, 14a and 15a, where the pore size is 1 µm, and the difference between these cases is only in the evaporator temperature setting. Other comparisons at the same value of the pore size are shown in Figures 13b, 14b and 15b for the pore size of 3 µm, and Figures 13c, 14c and 15c for the pore size of 7 µm. Due to the change in the setting of wall temperature, the resulting pressure difference is 0.3 kPa.

Determination of Mass Flow Rate of Working Fluid in the LHP Evaporator

An essential parameter needed for the analysis of the pressure rise is the mass flow rate of the working fluid passing through the evaporator. Due to the lack of the possibility of direct measurement of the mass flow rate, an attempt was made to estimate it for the tested conditions. By knowing the heat flux supplied to the evaporator casing and the pressure and temperature range measured at the inlet and outlet of the evaporator, the mass flow rate of the working fluid ($\dot{m}_f$) inside the LHP was determined based on the heat balance. The mass flow rate was determined from the ratio of the heat flux ($\dot{Q}$) supplied to the evaporator casing divided by the enthalpy (h) difference between the evaporator outlet and evaporator inlet. Enthalpy was determined using the data from the REFPROP 10.0 software:

$$\dot{m}_f = \frac{\dot{Q}}{h_{evap.out} - h_{evap.inl}} \quad (1)$$

The results of the calculations are presented in Table 5.

The data presented in the table indicate the correlation between the heat source temperature and pore size, which, in the case of water, causes an increase in the mass flow rate as the temperature of the heat source increases and increases when the radius of the pore increases. Ethanol seems to be the most optimal with the pore size equal to 7 μm.

The use of ethanol in the experimental rig with the same operating parameters caused the observation of several times higher values of the mass flow rate regardless of the heat source setting, comparing the results with the use of distilled water. Figure 16 presents the dependence of the heat flux relation in the function of the achieved flow rate of the working fluid. The figures show that the relationship is practically linear. The influence of the pore radius on the vapor outlet and the applied temperature is noticeable. Figure 17 presents the dependence of the applied heat flux on the pressure increase where the eac measurement comes from a different experiment. For water, the experimental points are arranged in a linear trend, while in ethanol, the trend line was not linear, which means that the measurements were burdened with greater error.

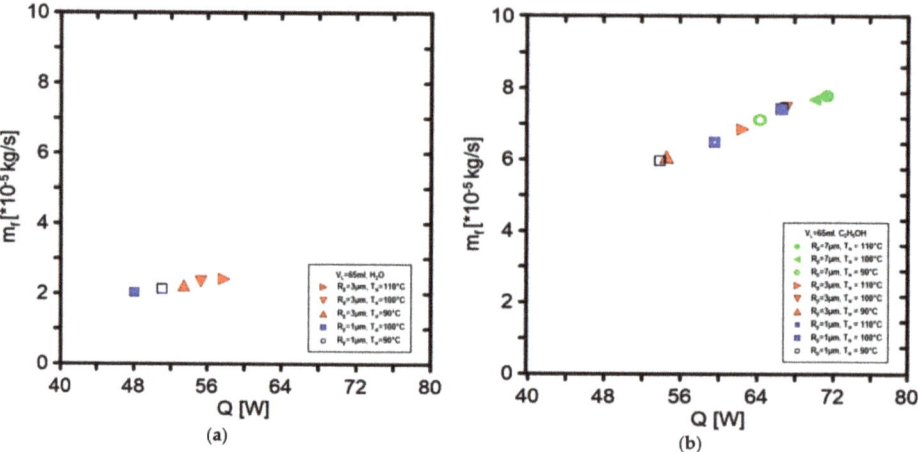

Figure 16. The mass flow rate in the function of applied heat flux for (**a**) water as a working fluid and (**b**) ethanol as a working fluid.

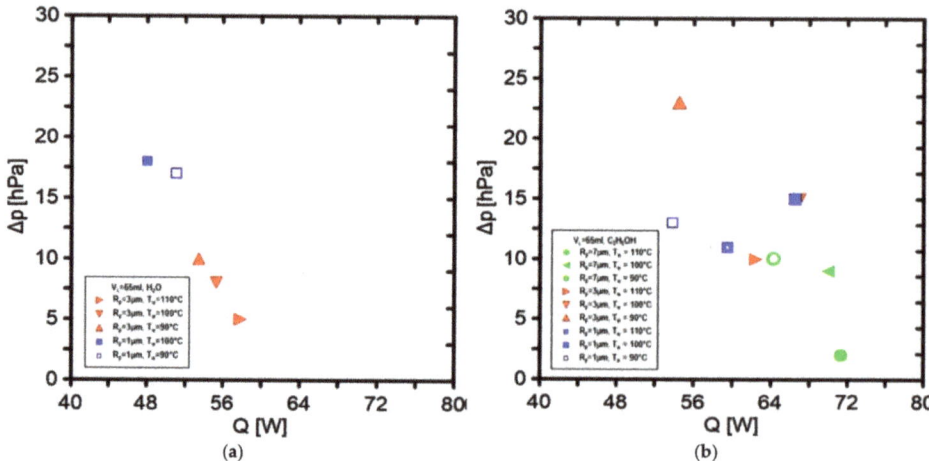

Figure 17. Pressure change in the evaporator in the function of applied heat flux for (**a**) water as a working fluid and (**b**) ethanol as a working fluid.

Table 5. Determination of the mass flow rate using the heat balance method.

Heater Temperature	Working Fluid	Pore Size	Heat Flux	Enthalpy-Evaporator Inlet	Enthalpy-Evaporator Outlet	Mass Flow Rate	P1–P2
[°C]	[-]	[μm]	[W]	[kJ/kg]	[kJ/kg]	[kg/s]	[kPa]
110	water	1	52.1	283.3	2642	2.08×10^{-5}	1.6
100	water	1	48.5	278	2638	2.03×10^{-5}	1.8
90	water	1	51.0	221.1	2621	2.13×10^{-5}	1.7
110	water	3	57.8	243.7	2640	2.41×10^{-5}	0.5
100	water	3	55.3	237.8	2608	2.33×10^{-5}	0.8
90	water	3	53.4	231.5	2633	2.23×10^{-5}	1.0
110	ethanol	1	59.5	337.6	1254	6.49×10^{-5}	1.1
100	ethanol	1	66.5	352.6	1247	7.4×10^{-5}	1.5
90	ethanol	1	53.8	331.3	1233	5.96×10^{-5}	1.3
110	ethanol	3	62.4	339.7	1249	6.86×10^{-5}	1.0
100	ethanol	3	67.0	342.6	1241	7.46×10^{-5}	1.5
90	ethanol	3	54.5	339.7	1236	6.08×10^{-5}	2.3
110	ethanol	7	71.3	359.2	1275	7.79×10^{-5}	0.2
100	ethanol	7	70.0	344.7	1255	7.69×10^{-5}	0.9
90	ethanol	7	64.2	342.3	1244	7.12×10^{-5}	1.0

4. Conclusions

The porous materials made of sintered stainless steel powder were tested as the evaporator wicks of the LHP. The experimental facility was assembled to study the efficiency of the various porous structures, characterised by the pore size of 1 μm, 3 μm and 7 μm for different working fluids. Two working fluids were tested, namely, water and ethanol.

The experiments indicate the crucial issue of adequate charge of installation with the working fluid. For both fluids considered, it was found that the pressure difference can reach up to 2.5 kPa for water as a working fluid and the pore size of 1 μm at the installation charge of 65 mL and 1.6 kPa in case the filling is 70 mL. This corresponds to 65% or 70% of the charge in installation. For other values of fillings, significantly lower values of pressure difference were obtained. In the case of ethanol, the results return a similar qualitative trend.

The respective values of mass flow rate were determined based on the energy balance for the evaporator and condenser separately. Good consistency of the results was obtained. The available mass flow rate is about 2.5 times higher in the case of ethanol than in the case of water at identical conditions.

Additionally, this research presents the feasibility of manufacturing inexpensive LHPs with filter medium as a wick material and its influence on the LHP's thermal performance.

Author Contributions: Conceptualisation, D.M. and K.B.; methodology, K.B.; software, K.B.; validation, D.M; formal analysis, K.B.; investigation, K.B.; resources, P.S.; data curation, P.S.; writing—original draft preparation, P.S.; writing—review and editing, P.S.; supervision, D.M. All authors have read and agreed to the published version of the manuscript.

Funding: This research received no external funding.

Institutional Review Board Statement: Not applicable.

Informed Consent Statement: Not applicable.

Conflicts of Interest: The authors declare no conflict of interest.

References

1. Pastukhov, V.G.; Maidanik, Y.F.; Vershinin, C.V.; Korukov, M.A. Miniature loop heat pipes for electronics cooling. *Appl. Therm. Eng.* **2003**, *23*, 1125–1135. [CrossRef]
2. Maydanik, Y.F. Loop heat pipes. *Appl. Therm. Eng.* **2005**, *25*, 635–657. [CrossRef]
3. Launay, S.; Sartre, V.; Bonjour, J. Parametric analysis of loop heat pipe operation: A literature review. *Int. J. Therm. Sci.* **2007**, *46*, 621–636. [CrossRef]
4. Chen, Y.; Groll, M.; Merz, R.; Maydanik, Y.F.; Vershinin, S.V. Steady-state and transient performance of a miniature loop heat pipe. *Int. J. Therm. Sci.* **2006**, *45*, 1084–1090. [CrossRef]
5. Ambirajan, A.; Adoni, A.A.; Vaidya, J.S.; Rajendran, A.; Kumar, D.; Dutta, P. Loop heat pipes: Review of fundamentals, operation, and design. *Heat Transfer. Eng.* **2012**, *33*, 387–405. [CrossRef]
6. Vasiliev, L.; Lossouarn, D.; Romestant, C.; Alexandre, A.; Bertin, Y.; Piatsiushyk, Y.; Romanenkov, V. Loop heat pipe for cooling of high-power electronic components. *Int. J. Heat Mass Transfer.* **2009**, *52*, 301–308. [CrossRef]
7. Bai, L.; Zhang, L.; Lin, G.; He, J.; Wen, D. Development of cryogenic loop heat pipes: A review and comparative analysis. *Appl. Therm. Eng.* **2015**, *89*, 180–191. [CrossRef]
8. Hoang, T.T. Cryogenic loop heat pipes for space and terrestrial applications. In Proceedings of the 4th International Conference and Exhibition on Mechanical & Aerospace Engineering, Orlando, FL, USA, 3–4 October 2016.
9. Esarte, B.; Bernardini, S.-J. Optimizing the design of a two-phase cooling system loop heat pipe: Wick manufacturing with the 3D selective laser melting printing technique and prototype testing. *Appl. Therm. Eng.* **2017**, *111*, 407–419. [CrossRef]
10. Jafari, D.; Wits, W.W.; Geurts, B.J. Metal 3D-printed wick structures for heat pipe application: Capillary performance analysis. *Appl. Therm. Eng.* **2018**, *143*, 403–414. [CrossRef]
11. Jafari, D.; Wits, W.W. The utilization of selective laser melting technology on heat transfer devices for thermal energy conversion applications: A Review. *Renew. Sustain. Energy Rev.* **2018**, *91*, 420–442. [CrossRef]
12. Szymanski, P. Recent Advances in Loop Heat Pipes with Flat Evaporator. *Entropy* **2021**, *23*, 1374. [CrossRef]
13. Richard, B.; Pellicone, D.; Anderson, W.G. Loop Heat Pipe Wick Fabrication via Additive Manufacturing. In Proceedings of the Joint 19th IHPC and 13th IHPS, Pisa, Italy, 10–14 June 2018.
14. Richard, B.; Anderson, B.; Chen, C.H.; Crawmer, J.; Augustine, M. Development of a 3D Printed Loop Heat Pipe. In Proceedings of the 49th International Conference on Environmental Systems, Boston, MA, USA, 7–11 July 2019.
15. Hu, Z.; Wang, D.; Xu, J.; Zhang, L. Development of a loop heat pipe with the 3D printed stainless steel wick in the application of thermal management. *Int. J. Heat. Mass. Transfer.* **2020**, *161*, 120258. [CrossRef]
16. Huang, B.-J.; Chuang, Y.-H.; Yang, P.-E. Low-cost manufacturing of loop heat pipe for commercial applications. *Appl. Therm. Eng.* **2017**, *126*, 1091–1097. [CrossRef]
17. Siedel, B.B. Analysis of Heat Transfer and Flow Patterns in a Loop Heat Pipe: Modelling by Analytical and Numerical Approaches and Experimental Observations. Ph.D. Thesis, INSA de Lyon, Campus de la Doua, France, 2014.
18. Mikielewicz, D.; Szymanski, P.; Wajs, J.; Mikielewicz, J.; Ihnatowicz, E. The new concept of capillary forces driven evaporator with application to waste heat recovery. In Proceedings of the VIII Minsk International Seminar: Heat Pipes, Heat Pumps, Refrigerators, Power Sources, Minsk, Belarus, 12–15 September 2011; Volume 1, pp. 316–323.
19. Mikielewicz, D.; Szymański, P.; Błauciak, K.; Wajs, J.; Mikielewicz, J.; Ihnatowicz, E. The new concept of capillary forces aided evaporator for application in Domestic Organic Rankine cycle. *Int. J. Heat Pipe Sci. Technol.* **2010**, *1*, 359–373. [CrossRef]
20. Mikielewicz, D.; Blauciak, K. Investigation of the influence of capillary effect on operation of the loop heat pipe. *Arch. Thermodyn.* **2014**, *35*, 59–80. [CrossRef]
21. Promotion materials by Tridelta Siperm GmbH, Dortmund. Available online: www.siperm.com (accessed on 18 November 2021).
22. Lemmon, E.W.; Bell, I.H.; Huber, M.L.; McLinden, M.O. *Refprop 10.0*; Applied Chemicals and Materials Division, National Institute of Standards and Technology (NIST) Software: Boulder, CO, USA, 2018.
23. Faghri, A. *Heat Pipe Science and Technology*; Taylor & Francis Inc: Washington, WA, USA, 1995.
24. Moffat, R.J. Describing the uncertainties in experimental results. *Exp. Therm. Fluid Sci.* **1988**, *1*, 3–17. [CrossRef]

Article

Influence of Holes Manufacture Technology on Perforated Plate Aerodynamics

Joanna Grzelak [1],* and Ryszard Szwaba [2]

1. Faculty of Mechanical Engineering and Ship Technology, Gdansk University of Technology, 11/12 Narutowicza, 80-233 Gdansk, Poland
2. Institute of Fluid Flow Machinery, Polish Academy of Sciences, Fiszera 14, 80-231 Gdansk, Poland; rssz@imp.gda.pl
* Correspondence: jj@imp.gda.pl

Abstract: Transpiration flow is a very important and still open subject in many technical applications. Perforated walls are useful for the purpose of "flow control", as well as for the cooling of walls and blades (effusive cooling) in gas turbines. We are still not able to include large numbers of holes in the numerical calculations and therefore we need physical models. Problems are related also to the quality of the holes in perforated plates. The present transpiration analysis concerns with experimental investigations of the air flow through perforated plates with microholes of 125 and 300 µm diameters. A good accordance of the results with other experiments, simulations and theory was obtained. The received results very clearly show that technology manufacturing of plate holes influences on their aerodynamic characteristics. It turned out that the quality of the plate microholes using laser technology and, consequently, the shape of the hole, can affect the flow losses. Therefore, this effect was investigated and the flow characteristics in both directions were measured, i.e., for two plate settings.

Keywords: perforated plates; laser micromachining; holes manufacture; flow direction

1. Introduction

Perforated plates have been used for many years in numerous industrial sectors and have been the subject of intensive research for decades [1–3]. Areas of application range are very different, from absorption of sound [4], separators and micro filters [5,6], heat exchangers [7,8], to a method of homogeneous flow generation [9] and a new cooling concept for blade turbines [10]. Transpiration flows are also used in flow steering devices to control the boundary layer separation, as well in passive shock wave control [11,12]. Laminar technology, which uses suction with perforated plates, is worth mentioning too. These aspects have been presented in works on laminar wing technology [12,13]. Perforated plates are the basis for microchannel cooling (or effusive cooling) of blades of gas turbines, where aerodynamics and heat transfer interaction take place. In this case, the use of plates allows the coolant to be distributed in a cooling layer with much better uniformity, as well as better transport of the coolant to the blade surface and creating better protection of the blade surface from hot gases [14].

The above examples of perforated plate applications, however, require the hole sizes in the order of 100 µm. Even a small area with a perforation of approximately 5% already requires thousands of holes. In this case, the use of drilling technology is unlikely to be an option due to the high costs involved. Additive manufacturing (3D printing) is also not a solution either, because with these hole sizes the powder material cakes inside the hole. The only technology which can cope with making such a large number of holes is laser technology.

The laser micromachining is most often carried out through direct illumination of the workpiece by the focused laser beam (so-called "direct writing") [15]. The lasers can operate in continuous-wave mode and also in pulsed mode. Pulse length for both lasers modes can be adjusted from 10 µs and 50 ms, with the exact range depending on the parameters used and laser model. During micromachining the laser pulses are focused on the material surface (e.g., metal). The absorbed energy is transported deep into the material in the process of thermal conductivity. The material located in the so-called heat-affected zone (HAZ) partially melts and then evaporates. However, it usually comes with the degradation of the processed material as a result of its damage around the micromachining spot. The consequence of such process are thermal effects as, e.g., deformation, sintering or discoloration of the edge of the material [16,17]. The disadvantages of laser technique include relatively low accuracy (on the level of several micrometers), and possibility of thermal damage of the workpiece material in HAZ.

Transpiration flows can be modeled in several ways. The first one that does not take into account the flow losses through the microholes is the isentropic model. Models supported by experimental data have been proposed, for example, in the works of Poll [18] or Inger and Babinsky [19,20]. They all relate to the air flow through cylindrical holes and include constants, which take into account the imperfections of the microholes. The models proposed by Inger and Babinsky contain the influence of viscosity and compressibility and also the shape of the opening of the hole. Numerical modeling of anisotropic drag for a perforated plate with an array of cylindrical holes was carried out in Youngmin and Young [21], but these results cannot be used for technical issues.

In experimental studies on perforated plates, it is usual to determine the characteristics of such a plate, i.e., the dependence of the pressure drop across the plate as a function of the mass flow rate. Such formulas are then successfully applied in numerical simulations [22]. The problem in experiments using perforated plates is less important that the hole channel performed by the laser is not actually cylindrical, as this feature can be taken into account in aerodynamic perforation. With hole sizes in the order of hundreds of micrometers, a serious problem from an aerodynamic point of view is that the hole diameter made on the laser side is slightly different from that on the exit side, causing the hole channel to become conical. This channel shape will affect the aerodynamic characteristics depending on which diameter, larger or smaller, is at the inlet of the channel. Therefore, the motivation, the aim for this work, was to show the influence of micro-hole technology on the aerodynamic characteristics of perforated plates. When using such plates in various applications, one should keep in mind that a given characteristic corresponds to a specific flow direction. With a change in the direction of the flow in the holes of a given plate, its characteristics change, as will be shown below in the experimental studies.

2. Experimental Setup

This paper concerns the flow direction effects across the perforated plates on their aerodynamic characteristics. The study of plate aerodynamic characteristics with microholes was carried out using general experimental data on a macro scale. In this regard, experimental investigations were performed using air as medium over two perforated plates with parameters displayed in Table 1, where D is a hole diameter, L—hole channel length, S—perforation, F—plate area.

Table 1. Plates parameters in the experiment.

Plate Name	D (mm)	L/D	S (%)	F (m^2)
K1	0.125	8.00	4.09	0.00283
K3	0.3	4.67	5.70	0.00283

Investigations have been performed for a range of Mach numbers from $Ma < 0.1$ up to choked condition of the flow in microholes, $M = 1$. The flow of compressible gas dynamics in channels has been studied considering different aspects of flow physics and can be found in various books and papers, e.g., the continuum approach [23], the molecular approach [24] and general analytical relations [25]. The basic theory necessary to obtain information from experimental data is outlined below. The equation of state of gas represents the relationship between the parameters of an ideal gas. When this equation is transformed, the speed of sound is defined as:

$$a = \sqrt{\kappa R T} = \sqrt{\frac{\kappa p}{\rho}} \quad (1)$$

R—gas constant, κ—specific heat constant, p—pressure, T—temperature, ρ—density.

Molecular properties related to the size of the continuum is another characteristic parameter of gas, namely the mean free path between the gas molecules translates into the average distance between them and is described as:

$$\lambda = \frac{k_\lambda \mu(T) \sqrt{2RT}}{p} \quad (2)$$

where $k_\lambda = \frac{\sqrt{\pi}}{2}$. Viscosity in this equation depends on the temperature and could be defined as:

$$\mu(T) = \left(\frac{T}{T_{ref}}\right)^\omega \quad (3)$$

k—thermal conductivity, μ—dynamic viscosity, p—pressure, ω—viscosity index, U—velocity.

The dimensionless parameters that determine the appropriate scales in the flows are defined by the Reynolds number Re and the Mach number Ma. The characteristic quantities which determine these parameters are based on the hole diameter and the average flow velocity, i.e., $Ma = \frac{U}{a}$, $Re = \frac{\rho U D}{\mu}$.

Experiments were conducted in the measurement section which is shown in Figure 1. The arrow in the figure indicates the direction of the flow, which is from left to right. The flow resulting from the pressure variance between the ambient and the pressure in the vacuum vessels downstream of the Valve 7. The air parameters in the laboratory space, i.e., the temperature of approximately 20–22 °C and the atmospheric pressure on a given measurement day correspond to the ambient conditions. The air flow starts from the ambient through the Flow Meter 1, further through the Control Valve 2, Compensation Chamber 3 to Frame 4. The tested perforated plate with micro holes is mounted in Frame 4. Behind the plate, air flows through Chamber 5, Valve 6 (the flow condition controller behind the micro-holes) and follows through cut-off Valve 7 into the vacuum vessels. A schematic diagram of the measurement system is shown in Figure 2.

During the experimental campaign, the following quantities were measured: mass flow rate, pressures, and temperature at the relevant points. Since the mass flow rate varied significantly depending on the pressure difference on both sides of the perforated plate, therefore depending on the mass flow range, this value was measured by means of three various laminar flow meters, (Alicate: 20, 100, 1500; Tucson, AZ, USA; SLPM; Standard Litre Per Minute at ambient condition, i.e., pressure of 1013 hPa and temperature of 298 K). The accuracy of mass flow rate measurement is ±0.01 SLPM of measured value. The stagnation parameters, i.e., pressure and temperature were measured in Compensation Chamber 3. Pressure was measured using a Dwyer Prandtl probe (Dwyer, Michigan City, ND, USA), Model 167-6, with a diameter of 1/8". A Kulite pressure transducer (Kulite, Leonia, NJ, USA) with accuracy of 0.1% (FS, full scale), i.e., ±1 hPa, was connected to this probe. Temperature was measured using a thermoelement (Czaki, Raszyn, Poland) with accuracy of 0.1 K. The pressure before and after the perforated plates was measured using a pressure scanner (MEAS, Hampton, VA, USA) with accuracy of 0.05% (FS), i.e., ±0.5 hPa.

The investigations of the flow through a porous plate were carried out under ambient conditions. The specified mass flow rate flows over area F, which induces a pressure difference $\Delta P = P_{in} - P_{out}$. The pressure P_{in} was equal to the atmospheric pressure P_0 and the pressure P_{out} downstream the plate was set using Valve 6. In this way, with appropriate manipulation of the control valve, the individual characteristics can be measured.

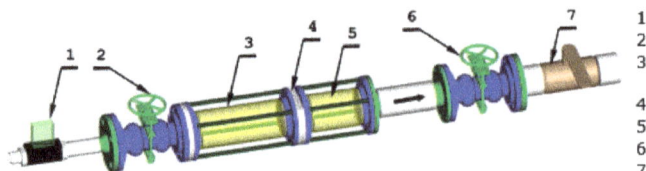

Figure 1. The test section view.

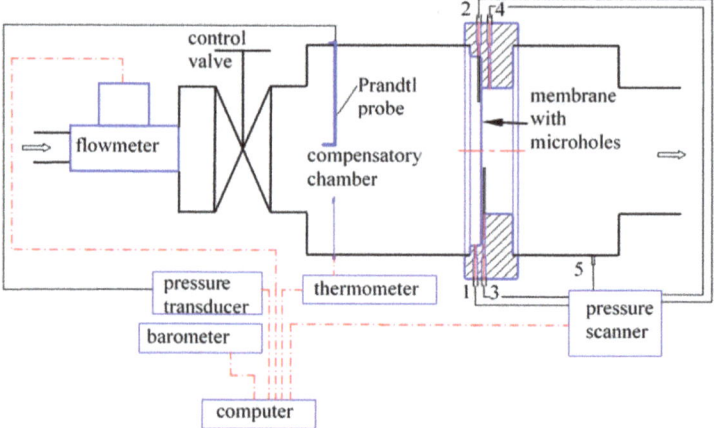

Figure 2. Arrangement of the measurement system.

A detailed zoomed view of the plate is shown in Figure 3. The diameter of perforated plates was 60 mm and they were manufactured from stainless steel. A microscopic view of K1 plate is shown in Figure 3. This is the view on the inlet side respect to the laser position and one can notice the degradation of the holes edge as a result of its damage around the laser spot. The accuracy of micro-hole diameter assessment is 10 μm and its width is 2 μm. The perforation values were given by the producer of plates. These values are presented in Table 1. The production porosity specification of plates K1 and K3 was a 4.1 and 5.7%, respectively. Actually, the aerodynamic porosity (the value that is measured during experiments) differs from that declared by the producer.

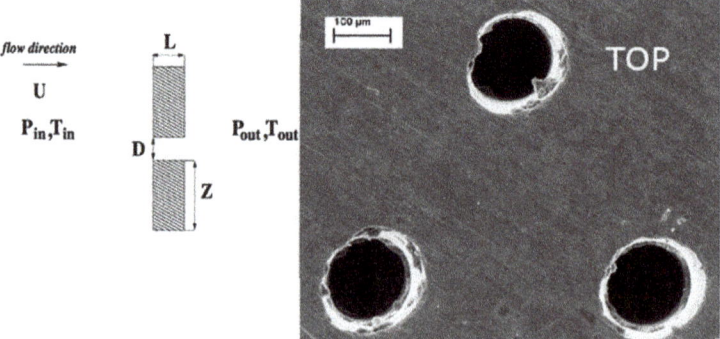

Figure 3. Schematic illustration of the perforated plate and the microscopic picture.

Aerodynamic porosity informs about what mass rate actually flows through the plate with a given cross-section (F) and at a given pressure difference. The differences between geometric and aerodynamic porosity are due to the inaccuracy of the hole manufacture and gas-dynamic effects. Hence, we use these two different definitions of the porosity. The porosity declared by the manufacturer results from the settings of the technological process, i.e., the diameter of the holes and their arrangement on a given surface. Aerodynamic porosity considers all these imperfections. The exact way of determining the aerodynamic porosity is presented in [26,27].

To verify the influence of the hole channel shape on the aerodynamic characteristics, the given plate was placed with the smaller hole opening at the inlet first (Figure 4a) and then reversed to obtain the bigger hole opening at the inlet (Figure 4b). The results for the latter plate setting are named as 'reversed' (rev). A 137 µm-diameter hole drilled in the plate represents the entrance of the laser beam (marked as TOP in Figures 3 and 4b), while 115 µm-diameter is the exit. The exit of the laser beam characterises with the smaller diameter and the conical shape of the hole. The lengths of the holes diameter were calculated by the computer program (Dynamic Studio 4.4.3). The quality of the picture, resolution, lighting, and contrast seen under the microscope is incomparably better compared to what can be seen in the picture in the article, so the measurement error is estimated at approx. ±2 px, which corresponds to approx. ±1 µm.

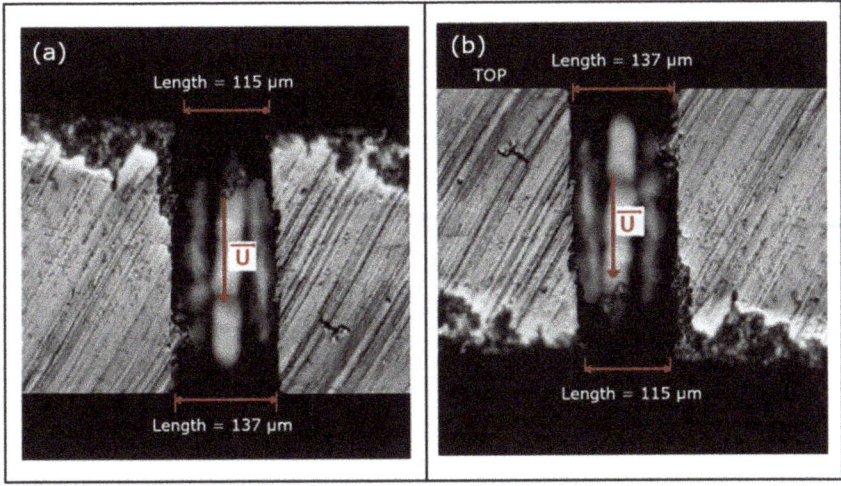

Figure 4. Shape of the hole channel with the velocity direction for the basic (**a**) and reversed (**b**) plate setting; plate K1.

Measurement uncertainties were estimated for the mass flow rate and the pressure difference, but they are very small and almost not visible on the picture (they are smaller than the chart points). To calculate the uncertainties of the above-mentioned quantities, exact differentials were used, containing such variables as pressure, temperature and air density. Average value of the obtained measurement errors for the mass flow rate are equal to 1.7%. High-accuracy pressure transducers (MEAS, Hampton, VA, USA) were used to measure the pressure, therefore the measurement errors for the pressure difference were smaller than 0.04%.

3. Experimental Results

The results presented in this section correspond to the basic plate setting (Figure 4a). The Reynolds number Re_{out} is based on the orifice's diameter and outlet parameters. The Mach number in the plate hole, Ma_h, is Reynolds numbers function (Figure 5). As can be noted, except for one point for K1, where $Re = 33$, there can be distinguished only one flow regime for Reynolds numbers higher than 50, which is linked to the turbulence transition. It can be also noticed the effects of compressibility in flow through the holes of perforated plate. The flow velocity (Ma_h—non-dimensional velocity coefficient) distribution at a high Reynolds number changes its linear character. These effects are stronger for the smaller hole diameter of the perforated plate.

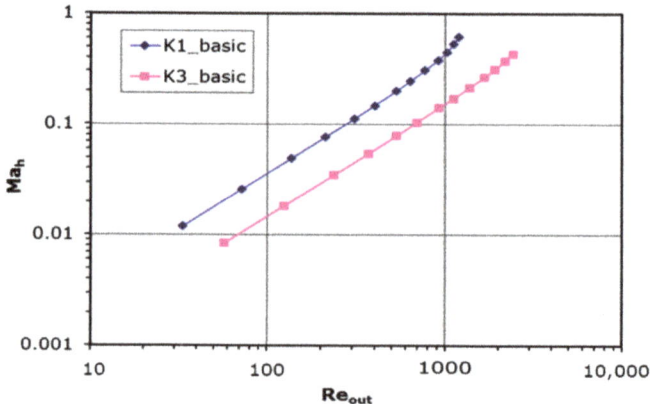

Figure 5. Mach number versus Reynolds number plot.

The mass flow rate Q is specified by the formula from the experimental studies in Reference [26]:

$$Q = FS\varrho_h U_h \qquad (4)$$

The mass flow rate for a single hole, $Q_h = Q/N$, as a function of Reynolds number is shown in Figure 6. Since the velocity resulting from the mass flow rate is a variable in the calculation of Reynolds number, the nature of this relationship is not unexpected. Therefore, the dependency between the mass flow rate Q_h and the Reynolds number is saw as almost linear (small deviation for high Reynolds numbers) and plotted in Figure 6. Higher mass flow rates for the same Reynolds number are obtained for the larger holes orifice in the whole range of the parameters.

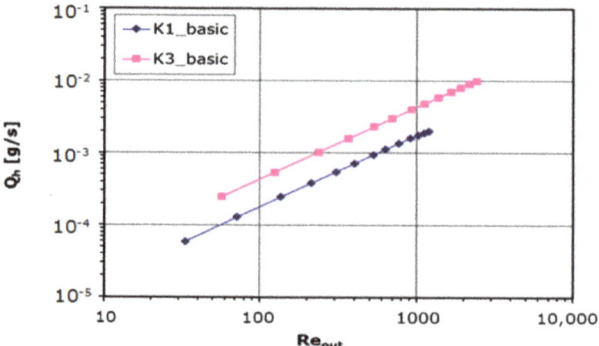

Figure 6. Mass flow rate versus Reynolds number plot.

3.1. Aerodynamics of Perforated Plates

The aerodynamics of perforated plates which was studied for years in many different experiments has been focused on plates with large holes. Examples that can be found in the literature concern for example the pressure loss coefficient over thickness of plate [28], turbulent flow over the perforated plate [9] and high Reynolds numbers ($Re > 10{,}000$) [29]. In this paragraph, the obtained results will be compared with the ones of two specific paper that considers various hole diameters in different kinds of plates [26,27]. Implementing the stagnation parameters to Equation (4), we obtain Equation (5).

$$Q = FS \frac{Ma_h}{\left(1 + \frac{\kappa-1}{2} Ma_h\right)^{\frac{\kappa+1}{2(\kappa-1)}}} P_0 \sqrt{\frac{\kappa}{RT_0}} \tag{5}$$

The experimental formula obtained in Reference [18] leads to:

$$Ma_h = 1.2 \left(\frac{\Delta P}{P_0}\right)^{0.55} \tag{6}$$

The curve resulting from Equation (6) is plotted together with experimental data in Figure 7. This plot shows that velocity coefficient is tending to unity with a pressure drop on both sides of the plate increases. The experimental data coincide very well with that model and the precise fitting coefficients are introduced in Table 2 beneath, where that relation is specified by the formula:

$$Ma_h = A \left(\frac{\Delta P}{P_0}\right)^B \tag{7}$$

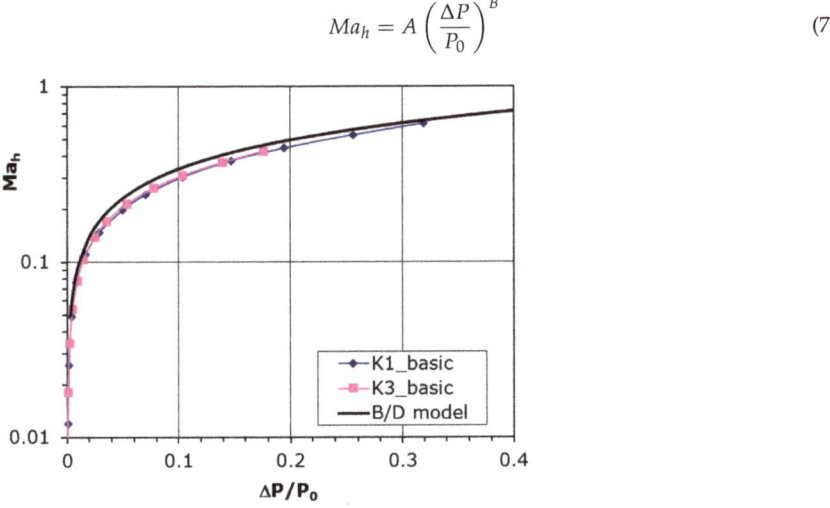

Figure 7. Comparison with other experiment for perforated plates in parallel flow, B/D model.

Table 2. Coefficients for the Equation (7) formula, Mach number dependence from pressure drop.

Plate	A	B
K1_basic	1.22	0.60
K3_basic	1.21	0.59

It can be seen that Coefficients A and B are very similar to that from Equation (6) and quite good compatibility with B/D model is achieved.

3.2. Characteristics Depending on the Flow Direction

Analysis connected with perforated plates usually involve cylindrical hole assumption, which is true when holes are made using drilling technology. In this experiment, laser technology was used to make the perforated plates, hence the hole channel has a shape closer to conical than to cylindrical. To verify if the shape of the hole channel really influences the aerodynamic characteristics, the given plate has been reversed to set the bigger hole opening at the inlet. The results for such plate setting are shown in Figure 8a (plate K1) and Figure 8b (plate K3). The pressure difference is a function of the mass flow rate in the hole, Q_h. It can be noted that for the same pressure difference the mass flow rate is smaller in the case of 'reversed' setting for both plates with various diameter. It is also seen that the Coefficients A and B from Equation (7) for the latter case also differ to those for the basic plate setting (Table 3 and Figure 9). The velocity in the hole increases slower with the pressure drop, which indicates a higher aerodynamic drag in the flow for the 'reversed' plate setting.

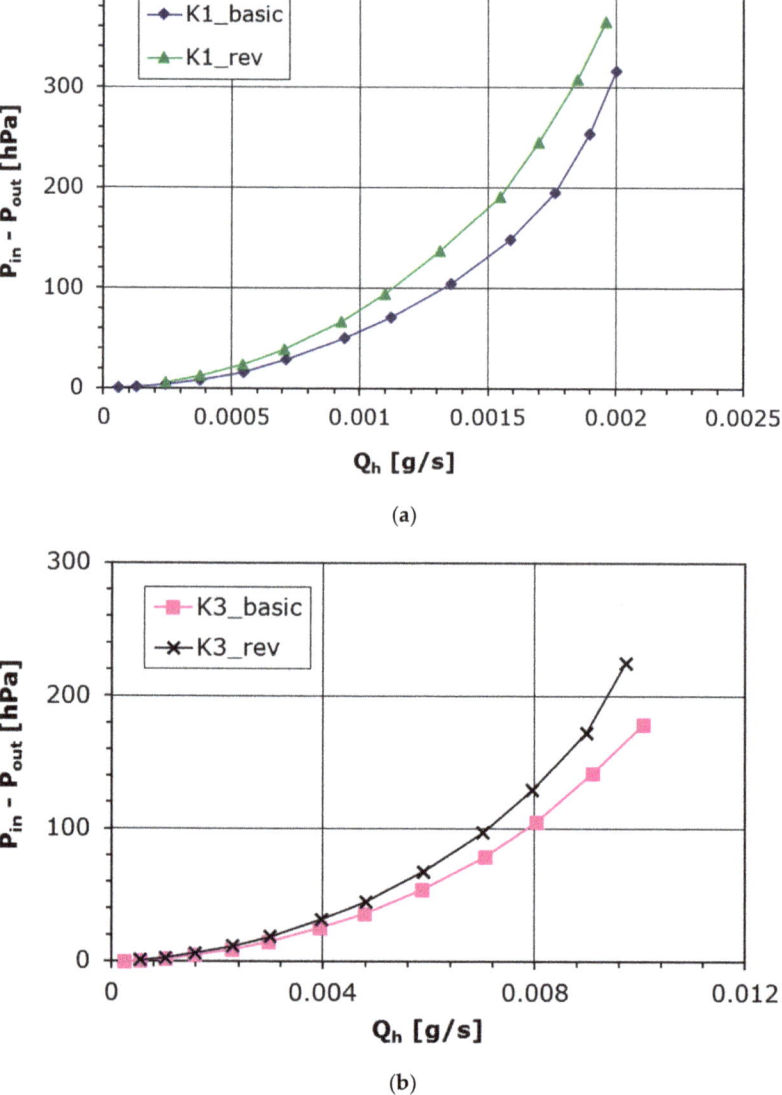

Figure 8. (**a**) Pressure difference as a function of the mass flow rate for basic and revised plate setting; Plate K1. (**b**) Pressure difference as a function of the mass flow rate for basic and revised plate setting; Plate K3.

Table 3. Coefficients for the Mach number pressure drop dependency for 'revised' plate setting.

Plate	A	B
K1_rev	1.06	0.60
K3_rev	1.04	0.58

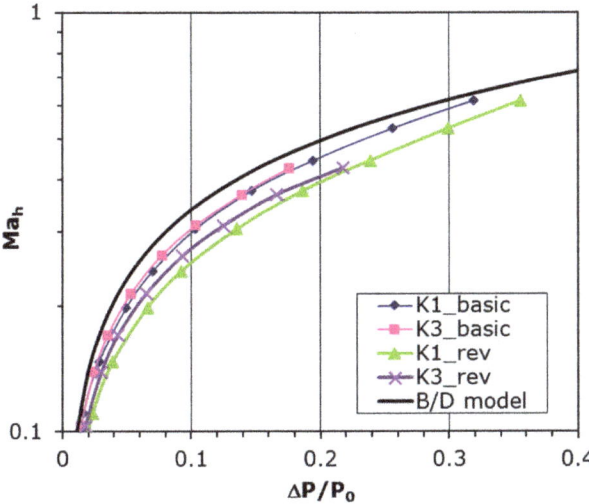

Figure 9. Comparison with B/D model for both plate settings and Plates K1, and K3.

The above results show how much the manufacturing of plate holes influences the aerodynamic characteristics. A lower pressure drop at the same flow rate was obtained for the basic setting, i.e., the smaller hole cross-section is located at the inlet of the flow through the plate. This behaviour is related to the shape of the hole, i.e., in subsonic flow with increasing cross-sectional area the flow velocity decreases and the pressure in the hole increases, which translates into a lower pressure drop across the plate compared to the reverse setting where the flow in the hole accelerates and the pressure loss increases.

The above results show how important the hole performance technology is in relation to the aerodynamic characteristics of perforated plates. If the laser technology is chosen, due to its influence on the shape of the holes as described in detail above, it is necessary before using perforated plates to accurately determine the direction of flow in the conical hole in the flow characteristics study of the plate. Further on, in the specific application of perforated plates, it is imperative that this flow direction is respected, i.e., the correct positioning of the plate in direction of the pressure drop in the flow.

4. Conclusions

This paper concerns the laser technology effects on the flow across the perforated plates and their aerodynamic characteristics. The measurements of the aerodynamic performance of the plate with microholes were carried out using the universal macroscale experimental data. Aerodynamic characteristics for two different plates with different perforations, hole diameters and channel lengths were presented. To verify the influence of the hole manufacture technology on the plates characteristics, two flow directions (basic and reversed plate setting) were studied.

The investigation showed that the laser technology of making microholes is not indifferent to the flow direction through the plate. For the same pressure difference the mass flow rate is smaller in case of 'reversed' setting for both plates, i.e., the smaller hole cross-section is located at the outlet of the flow through the plate. The experimental data were also compared with Equation (6) representing the B/D model. A good accordance was obtained, although the Coefficients A and B from Equation (7) are different for different plate settings. The received results very clearly show the technology manufacturing of plate holes influences on their aerodynamic characteristics. Perhaps a larger range of parameters (diameter, perforation) would show a greater difference. Nevertheless, this investigation requires continuation in the future.

In the case of perforated plate applications where the required holes diameter in the plate is below 0.3 mm, there is practically no choice of hole manufacturing technology, and we are forced to choose the laser technology. It is, therefore, very important that before using perforated panels, it is necessary to accurately determine the flow direction in the conical shape hole in the flow characteristics study of the plate, i.e., the same position of the plate in the direction of the pressure drop across the perforated plate both in the investigation and in the application.

Author Contributions: Conceptualization, R.S.; methodology, R.S.; software, R.S. and J.G.; validation, R.S. and J.G.; formal analysis, R.S..; investigation, J.G.; resources, R.S.; data curation, R.S. and J.G.; writing—original draft preparation, R.S. and J.G.; writing—review and editing, R.S.; visualization, R.S. and J.G.; supervision, R.S.; project administration, R.S. All authors have read and agreed to the published version of the manuscript.

Funding: This research received no external funding.

Institutional Review Board Statement: Not applicable.

Informed Consent Statement: Not applicable.

Data Availability Statement: The data supporting the reported results of this study can be made available from the corresponding author, upon request.

Conflicts of Interest: The authors declare no conflict of interest.

References

1. Budoff, M.; Zorumski, W. *Flow Resistance of Perforated Plates in Tangential Flow*; NASA Report No TM X-2361; NASA: Washington, DC, USA, 1971.
2. Zierep, J.; Bohning, R.; Doerffer, P. Experimental and Analytical Analysis of Perforated Plate Aerodynamics. *J. Therm. Sci.* **2003**, *12*, 193–197. [CrossRef]
3. Szwaba, R.; Ochrymiuk, T.; Lewandowski, T.; Czerwinska, J. Experimental Investigation of Microscale Effects in Perforated Plate Aerodynamics. *J. Fluids Eng. ASME* **2013**, *135*, 121104. [CrossRef]
4. Maa, D. Potentials of Micro Perforated Absorbers. *J. Acoust. Soc. Am.* **1975**, *104*, 2866–2868.
5. Van Rijn, C.; Van der Wekken, M.; Hijdam, W.; Elwenpoek, M. Deflection and Maximum Load of Microfiltration Membrane Sieves Made with Silicon Micromachining. *J. Microelectromech. Syst.* **1994**, *6*, 48–54. [CrossRef]
6. Yang, J.; Ho, C.; Yang, X.; Tai, Y. Micromachined Particle Filter with Low Power Dissipation. *J. Fluids Eng. ASME* **2001**, *123*, 899–908. [CrossRef]
7. Lahjomri, J.; Obarra, A. Hydrodynamic and Thermal Characteristics of Laminar Slip Flow Over a Horizontal Isothermal Flat Plate. *J. Heat Transf. ASME* **2013**, *135*, 021704. [CrossRef]
8. Babak, V.; Babak, T.; Kholpanov, L.; Malyusov, V. A Procedure for Calculating Matrix Heat Exchangers Formed from Perforated Plates. *J. Eng. Phys.* **1986**, *50*, 330–335. [CrossRef]
9. Liu, R.; Ting, D. Turbulent Flow Downstream of a Perforated Plate Sharp-Edged Orifice Versus Finite-Thickness Holes. *J. Fluids Eng. ASME* **2007**, *129*, 1164–1171. [CrossRef]
10. Bunker, R. Gas Turbine Heat Transfer Ten Remaining Hot Gas Path Challenges. *J. Turbomach. ASME* **2007**, *129*, 193–201. [CrossRef]
11. Babinsky, H.; Ogawa, H. SBLI control for wings and inlets. *Shock Waves* **2008**, *18*, 89. [CrossRef]
12. Streit, T. DLR natural and hybrid transonic laminar wing design incorporating new methodologies. *Aeronaut. J.* **2015**, *119*, 1303–1326. [CrossRef]
13. Beck, N.; Landa, T.; Seitz, A.; Boermans, L.; Li, Y.; Radespiel, R. Drag Reduction by Laminar Flow Control. *Energies* **2018**, *11*, 252. [CrossRef]
14. Gau, C.; Huang, W. Effect of Weak Swirling Flow on Film Cooling Performance. *J. Turbomach. ASME* **1990**, *112*, 786–791. [CrossRef]
15. Piqué, A.; Chrisey, D.B. *Direct-Write Technologies for Rapid Prototyping Applications: Sensors*; Academic Press: San Diego, CA, USA, 2002; p. 385.
16. Garasz, K.; Tański, M.; Kocik, M.; Barbucha, R. Precise micromachining of materials using femtosecond laser pulses. In Proceedings of the SPIE—The International Society for Optical Engineering, Barcelona, Spain, 4–6 May 2015; Volume 9520. [CrossRef]

17. Mendes, M.; Sarrafi, R.; Schoenly, J.; Vengemert, R. Fiber laser micromachining in high-volume manufacturing, Industrial Laser Solutions. In Proceedings of the 15th International Symposium on Laser Precision Microfabrication, LPM2014, Vilnius, Lithuania, 17–20 June 2014.
18. Doerffer, P.; Bohning, R. Modelling of Perforated Plate Aerodynamics Performance. *Aerosp. Sci. Technol.* **2000**, *4*, 525–534. [CrossRef]
19. Inger, G.; Babinsky, H. Viscous Compressible Flow Across a Hole in a Plate. *J. Aircr.* **2000**, *37*, 1028–1032. [CrossRef]
20. Galluzzo, P.; Inger, G.; Babinsky, H. Viscous Compressible Flow Through a Hole in a Plate, Including Entry Effects. *J. Aircr.* **2002**, *39*, 516–518. [CrossRef]
21. Youngmin, B.; Young, I.K. Numerical modeling of anisotropic drag for a perforated plate with cylindrical holes. *Chem. Eng. Sci.* **2016**, *149*, 78–87.
22. Doerffer, P.; Szulc, O. Shock Wave Strength Reduction by Passive Control Using Perforated Plates. *J. Therm. Sci.* **2007**, *16*, 97–104. [CrossRef]
23. Howarth, L. *Modern Developments in Fluid Dynamics High Speed Flow*; Clarendon: Oxford, UK, 1953; Volume 2.
24. Bird, G. *Molecular Gas Dynamics and the Direct Simulation of Gas Flows*; Oxford University Press: New York, NY, USA, 1994.
25. White, F. *Fluid Mechanics*; McGraw-Hill: New York, NY, USA, 1998.
26. Poll, D.; Danks, M. *The Aerodynamic Performance of Laser Drilled Sheets*; Technical Report 92-02-028; University of Manchester, Aerospace Systems and Technologies: Stanley, UK, 1992.
27. Grzelak, J.; Doerffer, P.; Lewandowski, T. The efficiency of transpiration flow through perforated plate. *Aerosp. Sci. Technol.* **2021**, *110*, 106494. [CrossRef]
28. Gan, G.; Riffat, S. Pressure Loss Characteristics of Orifice and Perforated Plates. *Exp. Therm. Fluid Sci.* **1997**, *14*, 160–165. [CrossRef]
29. Cuhadaroglu, B.; Akansu, Y.; Turhal, A. An Experimental Study on the Effects of Uniform Injection Through One Perforated Surface of a Square Cylinder on Some Aerodynamic Parameters. *Exp. Therm. Fluid Sci.* **2007**, *31*, 909–915. [CrossRef]

Article

Design Concepts and Performance Characterization of Heat Pipe Wick Structures by LPBF Additive Manufacturing

Konstantin Kappe [1,*], Michael Bihler [1], Katharina Morawietz [2], Philipp P. C. Hügenell [2], Aron Pfaff [1] and Klaus Hoschke [1]

1. Fraunhofer Institute for High-Speed Dynamics (EMI), Ernst-Zermelo-Str. 4, 79104 Freiburg, Germany
2. Fraunhofer Institute for Solar Energy Systems (ISE), Heidenhofstraße 2, 79110 Freiburg, Germany
* Correspondence: konstantin.kappe@emi.fraunhofer.de

Abstract: Additive manufacturing offers a wide range of possibilities for the design and optimization of lightweight and application-tailored structures. The great design freedom of the Laser Powder Bed Fusion (LPBF) manufacturing process enables new design and production concepts for heat pipes and their internal wick structures, using various metallic materials. This allows an increase in heat pipe performance and a direct integration into complex load-bearing structures. An important influencing factor on the heat pipe performance is the internal wick structures. The complex and filigree geometry of such structures is challenging in regards to providing high manufacturing quality at a small scale and varying orientations during the printing process. In this work, new wick concepts have been developed, where the design was either determined by the geometrical parameters, the process parameters, or their combination. The wick samples were additively manufactured with LPBF technology using the lightweight aluminum alloy Scalmalloy®. The influence of the process parameters, geometrical design, and printing direction was investigated by optical microscopy, and the characteristic wick performance parameters were determined by porosimetry and rate-of-rise measurements. They showed promising results for various novel wick concepts and indicated that additive manufacturing could be a powerful manufacturing method to further increase the performance and flexibility of heat pipes.

Keywords: additive manufacturing; heat pipes; laser powder bed fusion; wick structures; heat pipe performance

Citation: Kappe, K.; Bihler, M.; Morawietz, K.; Hügenell, P.P.C.; Pfaff, A.; Hoschke, K. Design Concepts and Performance Characterization of Heat Pipe Wick Structures by LPBF Additive Manufacturing. *Materials* 2022, *15*, 8930. https://doi.org/10.3390/ma15248930

Academic Editor: Anatoliy Pavlenko

Received: 18 November 2022
Accepted: 10 December 2022
Published: 14 December 2022

Publisher's Note: MDPI stays neutral with regard to jurisdictional claims in published maps and institutional affiliations.

Copyright: © 2022 by the authors. Licensee MDPI, Basel, Switzerland. This article is an open access article distributed under the terms and conditions of the Creative Commons Attribution (CC BY) license (https://creativecommons.org/licenses/by/4.0/).

1. Introduction

The high energy and packing densities in certain industries, such as the automotive industry with its electrically powered cars, high-tech and computer hardware industries, and the aviation and space flight industries, drive the need for effective and reliable heat transfer. With this, heat pipes are gaining relevance as they have the ability to effectively transfer heat over a large distance [1]. The vital elements of the heat pipes are the inner wick structures [2,3]. They enable fluid transfer through capillary action and thus significantly determine their performance. However, the currently available uniform wick structures, such as grooved or sintered structures, suffer from conventional manufacturing constraints [4–6]. These limit their size and shape and thus the performance of the heat pipe [7,8]. Therefore, grooved and sintered shape type structures have recently been combined within a wick design in order to achieve better properties [7]. However, with conventional techniques, this requires increasingly complex manufacturing routes.

Current progress in the additive manufacturing of metals, in particular Laser Powder Bed Fusion (LPBF), offers a promising manufacturing process, as it allows great design freedom in terms of shape, geometry, and material properties [9,10]. These advances have created new possibilities for the design of heat-transferring devices [11,12]. Specifically, the additive manufacturing of heat pipes can enable new design possibilities and improved performance. New and unusual heat pipe geometries, as well as their direct integration into functional components, can be achieved [13–16]. Furthermore, the manufacturing conditions are customizable to a large degree, e.g., by adapting the laser exposure, such that the material properties of the wick structure could also be advanced and tailored on a small scale. This allows the inner wick structures to be directly determined by the geometric design [8,11,12,17] or the process parameters [18].

Ameli et al. [8] used LPBF with aluminum powder to manufacture cubic porous wick samples by forming octahedral unit cells with a regular and random distribution. These samples were characterized by their permeability and porosity and achieved comparable values to conventionally sintered wicks. Jafari et al. [19] used analysis methods introduced by Holley and Faghri [20] to conduct a detailed analysis of similar additively manufactured porous wicks from stainless steel and recorded significantly higher permeabilities and wick performance parameters at similar porosities. In further work, Jafari et al. [18] utilized the process parameters of the additive manufacturing process to create a porous bulk material using stainless steel suitable for heat pipe wick structures. They investigated the capillary performance, porosity, and pore radius of the printed samples, measuring an average porosity of 2.5–43% and pore radii of 9–23 µm. An application of an additively manufactured primary wick of a Loop Heat Pipe (LHP) was tested by Esarte et al. [6]. They developed geometrically determined porous structures and manufactured them using stainless steel powder. The heat transfer test showed a 10% increase compared to a LHP with a conventionally manufactured primary wick. Chang et al. [21] manufactured a flat aluminum heat pipe with grooved wick structures. The authors suggested that the rough surface of the additive manufacturing process further increases the capillary performance. In a previous case study, Kappe et al. designed and integrated various concepts of heat pipes into complex test structures [15]. The test samples were additively manufactured from Scalmalloy® by LPBF and examined in thermal experiments. The comparison revealed multiple challenges of the different structural concepts and manufacturing characteristics. However, the general feasibility was demonstrated and the first promising integrated heat pipes could be fabricated.

While additive manufacturing has the potential to push the boundaries of possibilities for heat pipe design and has motivated a range of research work, it still has some challenges and limitations:

- Metals with low density and high thermal conductance, such as aluminum, are favored for heat pipes, as they enable a lightweight design and increase the heat dissipation over the structure [1]. However, in LPBF, the properties of aluminum alloys pose a challenge for manufacturing filigree wick structures as the resolution is typically decreased in comparison to other materials. Many studies, therefore, are still confined to low thermal conductance materials such as steel or titanium alloys.
- The characteristics of LPBF depend strongly on the direction of printing. The presented results and improved wick characteristics have mostly been based on samples manufactured in the ideal, vertical orientation, where no overhang is introduced by the cylindrical shape. For designs with a more complex shape or the integration into a structural component, the printing orientation can vary for each section of the heat pipe. With increasing deviation from the ideal position, multiple problems arise, such as the necessity of support structures and specialized laser exposure strategies such as downskin.
- The characterization of the wick performance differs depending on the test method used, which makes comparability between different studies using additive manufacturing and conventional designs difficult.

This work aimed to develop different concepts for additively manufactured wick structures to achieve better wick performance and manufacturability even with unfavorable printing directions and materials. Different characterization methods were used to investigate these influences. The wick concepts presented in this paper were designed for an exemplary application in the optical bench of a nanosatellite [22–24]. However, the design and integration of the heat pipe still require further investigation. Grooved wick structures with different geometries, porosities, and hybrid concepts were designed and manufactured by LPBF using Scalmalloy® metallic powder. In addition, the influence of the manufacturing parameters and the printing orientation was investigated. The concepts were examined by means of optical microscopy to evaluate the manufacturing quality. With mercury intrusion porosimetry and helium pycnometry combined with rate-of-rise measurements, the wick performance of the manufactured samples was characterized. The results of the different measurement methods were compared to identify deviations and possible limits of the setups.

2. Design of the Wick Samples

The different concepts for additively manufactured wick structures can be classified into three different categories, whereby the design of the wick structure is determined by

- its geometric parameters,
- the LPBF process parameters,
- both the geometric parameters and process parameters.

For the geometrically defined wick structures, different groove shapes, namely triangular, rectangular, trapezoidal, arterial, and sloped grooves were selected. Figure 1a shows the CAD models of the grooved wicks and their characteristic geometrical parameters. A parameter study was carried out to provide initial statements on the required characteristic dimensions. This was based on a detailed theory described in [25], which refers to a one-dimensional model. An optimization for the respective case was conducted by calculating various parameter combinations. The characteristic dimensions, as seen in Figures 1a and 2a, were outputs of the parameter study and are summarized in Table 1. The sloped grooves shown in Figure 1b were developed to compensate for some of the problems arising from a horizontal printing orientation. The walls between the groove had a varying angle to the horizontal direction, ranging from 90° at the bottom to 45° in the middle. In this way, overhanging surfaces with angles under 45° were avoided. At the same time, the walls were designed to be fabricated with a single laser track each to minimize unwanted fusing of unmelted powder.

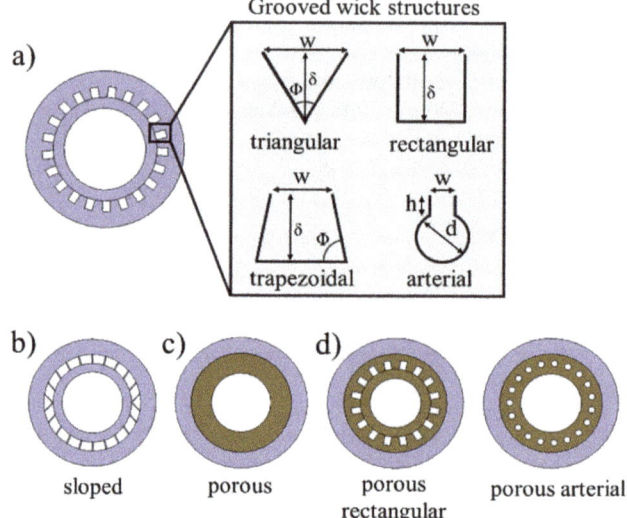

Figure 1. Different wick structures: (**a**) geometrically determined grooved wick structures and characteristic dimensions, (**b**) sloped grooved wick structure, (**c**) porous wick structure, and (**d**) hybrid wick structures.

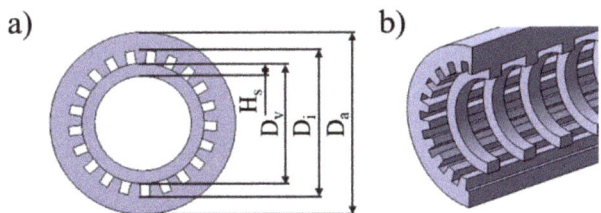

Figure 2. (a) Characteristic dimensions of wick sample and (b) support rings inside the grooved wick structures.

Table 1. Grooved wick dimensions.

Groove Type	Dimension	Value
general	outer ⌀ D_o [mm]	10.0
	inner ⌀ D_i [mm]	8.0
	vapor channel ⌀ D_v [mm]	6.6
	groove height δ [mm]	0.7
	sample length L [mm]	50.0
triangular	groove width w [mm]	0.5
	opening angle Φ	39.3°
rectangular	groove width w [mm]	0.5
trapezoidal	groove width w [mm]	0.5
	opening angle Φ	70°
arterial	groove width w [mm]	0.25
	artery ⌀ d [mm]	0.5
	channel height H [mm]	0.25

The geometry of the porous wick structure shown in Figure 1c was defined by the LPBF process parameters and consisted of a cylindrical volume element inside the solid outer wall. At first, the pores were deliberately formed by the manufacturing process. For the fabrication of porous structures, three main parameters were identified from previous experience and parameter studies: the laser power P, the laser track scan speed v, and the hatch distance h between the laser tracks. These parameters were varied to produce different porous samples. Multiple approaches exist for obtaining a single parameter to compare the processing conditions; see [26]. Here, the specific energy density e was used, which is defined as:

$$e = P/(v \times h) \qquad (1)$$

In order to create a fluid loop in the heat pipe, a porosity ε of approximately 45%, comparable to conventionally sintered wicks, and a permeability $K > 1 \times 10^{-10}$ were determined by the parameter study.

The porous rectangular and porous arterial wick as shown in Figure 1d aimed to increase the wick permeability by implementing paths with low flow resistance into the wick. At the same time, the porous sections should lead to a high capillary force. As there are no theoretical models to predict their performance, the same manufacturing parameters were applied and the same porosity values were targeted.

The intended minimum wall thickness of $t \geq 1.0$ mm was recommended in the material data sheet [27] and the guide value of overhang angles $\delta \leq 45°$ could not be met for all designs and printing orientations. To test the boundaries of manufacturability for structures deviating from these guidelines, non-standard manufacturing parameters were used based on experiences from previous parameter studies. Furthermore, inside the vapor channel of the grooved and hybrid wicks, support rings were implemented, as can be seen in Figure 2b. They supported the upper section of the wick when printing horizontally to reduce related problems in manufacturability. The depth of the rings was 1.0 mm with a height H_s of 0.7 mm, spaced out with gaps of 2 mm along the samples. This resulted in a 33% coverage of the liquid-vapor interface in the vapor channel, which might have negatively influenced the fluid loop but increased the manufacturability.

3. Additive Manufacturing by LPBF and Specimen Preparation

The wick concepts were additively manufactured using an EOS M 400 LBPF machine. The specimens were made of the aluminum alloy Scalmalloy® with a particle size distribution of D10 = 2010 μm, D50 = 3610 μm, and D90 = 5373 μm measured with a Microtrac MRB Camsizer X2. A layer thickness of 60 μm for all samples was specified. To provide an inert atmosphere, nitrogen was used. A rotating stripe scanning strategy with scan vectors rotated for every layer was utilized. In addition, a single contour exposure was used for the geometrically determined wick structures, while no contour exposure was used for the porous samples. For the manufacturing of the grooved wick concepts, five different manufacturing contour parameter sets were chosen, as shown in Table 2. The contour laser track parameter draws the outline and therefore has the biggest influence on the surface quality and the resolution of surface features. Parameters P1 and P2 were process contour parameters commonly used for printing bulk material of Scalmalloy® at Fraunhofer EMI, whereby parameters 3–5 were adapted for the printing of the filigree grooves.

Table 2. Contour parameters for grooved wicks, with parameters P1 and P2 as contour parameters commonly used for bulk material with different surface roughnesses, and parameters P3, P4, and P5 adapted for printing filigree structures.

Parameter	$P_{Contour}$ [W]	$v_{Contour}$ [mm/s]
P1	600	400
P2	900	3000
P3	600	3000
P4	100	250
P5	300	500

The parameters for the porous wick samples were chosen based on the results from preliminary porosity measurements. The influence of the energy density and hatch was visible in the density of the wicks. The porosities determined with the weighing method ranged between 38% and 60% and therefore achieved the target porosity of 30% to 60%. Here, the main influence was the energy density, with the lowest energy density reaching the highest porosity. Simultaneously, the porosity decreased with the hatch distance. Three different energy densities with different laser power and relatively high hatch distances between 0.46 and 0.67 mm were picked based on two criteria: high porosity and small pores. The porosity of the sample should be high to increase the permeability of the wick. Simultaneously, the pores should be small to increase the capillary pressure as the driving force of the fluid loop. The porous wicks were printed using the parameters shown in Table 3.

Table 3. Printing parameters for porous wicks.

Parameter	P1	P2	P3
P [W]	335	600	500
v [mm/s]	2950	3000	2000
h [mm]	0.46	0.64	0.67
e [J/mm^2]	0.25	0.31	0.37
ε [%]	59.6	45.8	38.4

The removal of the residual powder after the printing process was done in two steps. By cleaning with compressed air, the majority of the remaining powder was removed. In a final cleaning step, the samples were dispersed in an ethanol bath and treated in an ultrasonic bath. This way, the residual powder, especially in the fine pores of the porous wicks, could be further reduced. However, single powder residues could still be detected. For all samples, a specimen of 10 mm in length for the optical microscopy was cut with an abrasive wet cutting machine. This cutting process created a planar surface on the wick sample, which was used as a contact point with the liquid in the rate-of-rise measurements. For the samples examined with the mercury intrusion method, two additional 10 mm specimens were prepared.

The specimens for the optical microscopy were further processed. Firstly, they were cold-mounted in epoxy resin. Then they were ground and subsequently polished. Polishing cloths with different grain sizes were used to increase the surface quality. This process was cooled with a lubricant and a diamond spray was used as a polishing agent.

4. Experimental Setup

4.1. Microscopy and Preliminary Measurements

To study the microstructure of the wick samples, a Zeiss Axio Imager.Z2m optical microscope was used. The microscope was equipped with a camera to produce digital images. The microstructure features, such as the characteristic dimensions of the grooved wicks and the pore diameter of the porous wicks, were measured, as seen Figure 3.

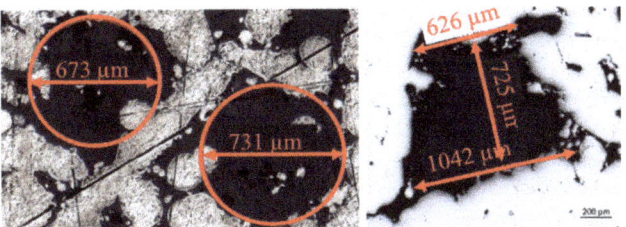

Figure 3. Exemplary measurements of pores (**left**) and trapezoidal grooves (**right**).

For a preliminary assessment of the porosity of the porous wicks, cubical samples were manufactured with identical process parameters. Their porosity was determined by the following relation [28]:

$$\varepsilon = 1 - \frac{\rho_a}{\rho_t} \tag{2}$$

where ρ_t was the bulk material density. The apparent density was $\rho_a = m/V$, where the mass m was determined by weighing and the volume V by measuring the outer dimensions of the cube samples.

4.2. Mercury Intrusion Porosimetry

The mercury intrusion porosimetry was carried out with a Quantachrome Poremaster 60. It measures pore sizes between 3.6 nm and 1100 μm by intruding the sample with non-wetting mercury [29]. The evaluated sample was placed in a glass cell. The latter was connected to a pressure chamber and evacuated. With the following intrusion of mercury, the pressure increased continuously. Simultaneously, the volume change was recorded at the stem of the measurement assembly. This dataset was then further analyzed by applying the Washburn equation, first formulated in [29,30]:

$$p = \frac{-2\sigma * \cos\theta}{r_p} \tag{3}$$

wherein σ represented the surface tension of the liquid, θ the contact angle between the liquid and the solid, which was assumed to be 140° for the combination of mercury and aluminum, and r_p as the cylindrical pore radius. By differentiation of (3) with the assumption of constancy of σ and θ, the volume pore size distribution $D_v(r_p)$ was determined to be:

$$D_v(r_p) = \frac{p}{r_p}\frac{dV}{dp} \tag{4}$$

A series of $\Delta V/\Delta p$ measured with the mercury intrusion method as a cumulative curve was then reduced to a distribution of pore volume per radius interval. Furthermore, the permeability of the wick structure could be determined as a function of the pore diameters and porosity [31]:

$$K = \frac{\varepsilon_{wick} d_p^2}{16\tau} \quad (5)$$

with d_p being the mean pore diameter gained from analyzing the pore size distribution. The porosity was calculated to be

$$\varepsilon_{wick} = V_{intruded}/(A_{wick} * l) \quad (6)$$

with $V_{intruded}$ being the total volume of the intruded mercury and A_{wick} the wick area of the sample. The measured porous area in the polished cut image was defined as the wick area and l was the sample length. The effective pore tortuosity τ was a modeled measure for the deviation of the pore shape of the straight cylindrical capillaries and straight diffusion paths. This was also determined in the mercury intrusion measurement as described in [32].

4.3. Helium Pycnometry

The Quantachrome MICRO-ULTRAPYC 1200e was used for further analysis of the density of the printed samples. A helium pycnometer was used to determine the true density of porous samples with open pores. This was based on the constant volume principle and included a sample and a reference chamber with a known volume V_R, which was connected via an initially closed transfer valve. The sample was placed in the sample chamber with the volume V_C, which was then filled with helium by increasing the pressure. The ultimate pressure p_1 was recorded and the transfer valve to the reference chamber opened. After the pressure equalization, the pressure p_2 was recorded. With these values, the sample volume V_P could be calculated by [33]:

$$V_P = V_c + \frac{V_R}{1 - (p_1/p_2)} \quad (7)$$

With the sample mass m and the apparent density $æ_a$, the true density $æ_t = m/V_P$ was put into $\varepsilon_{th} = 1 - \frac{\rho_a}{\rho_t}$ to receive the theoretical porosity ε_{th}.

4.4. Rate-of-Rise Experiment

For the rate-of-rise experiment, the Sartorius R160P analytical balance was used. For the experimental setup, a beaker with a platform, a frame to hang the sample, and a thermometer were used, as seen in Figure 4. The wick sample was hung into the frame with a sample holder. The latter had an opening with decreasing diameter in which to wedge the wick sample. By encasing the whole setup, the measurement error of the scale by circulating air, and the vaporization when using volatile working fluids were reduced. The temperature of the liquid and the ambient temperature were recorded with the thermometer to derive the fluid properties. Before each experiment, the balance was calibrated with an internal calibration weight.

As the working liquid, distilled water at 22 °C was used for the majority of the rate-of-rise experiments. It should be noted that distilled water is not the desired working liquid for the heat pipe application. It has been found that acetone in combination with Scalmalloy® is a more viable working liquid. To bring the wick sample in contact with the liquid, a syringe was used to fill the reservoir. The filling was stopped at the moment of contact, which was observed when the liquid meniscus enclosed the bottom of the sample. The experiment was stopped after 60 s, as the equilibrium height was reached for all samples after this period. The output of the rate-of-rise measurements was the capillary pumping mass as a function of time. To derive the wick performance from this data, a linear fitting approach as commonly used in studies [7,19,34], was utilized. With the following assumptions, the liquid rising process can be described [20] as:

- One-dimensional and steady-state laminar flow in the wick,
- uniform saturation with liquid along the wetted length,
- no initial and entry effects in the liquid reservoir,
- and evaporation of the liquid is neglected as the closed space by the glass cover minimizes the evaporation of the test liquid.

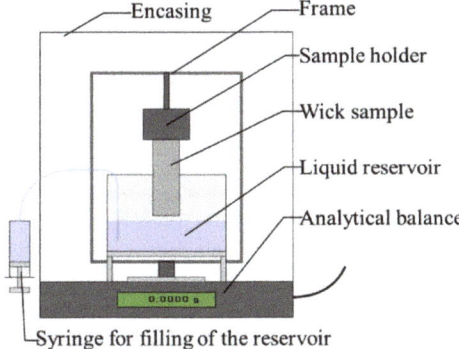

Figure 4. Rate-of-rise measurement setup.

For the rise of a liquid in the wick structure, the momentum balance of the capillary pressure Δp_{cap}, the pressure loss according to friction Δp_f, and the hydrostatic pressure Δp_h in the wetted height of the sample was:

$$\Delta p_{cap} = \Delta p_f + \Delta p_h \tag{8}$$

The description of the capillary pressure by the Laplace–Young equation, which is directly related to the Washburn equation [34] and the viscous friction by Darcy's law, led to:

$$\frac{2\sigma}{r_{eff}} = \frac{\mu \varepsilon}{K} h \frac{dh}{dt} + \rho g h \tag{9}$$

Here r_{eff} was the effective pore radius with $r_{eff} = r_p / \cos \theta$, μ the dynamic viscosity, g was the gravitational acceleration, h the capillary rise height, and dh/dt the capillary rise velocity. For the porosity ε, the measured values from the helium pycnometry were used where available. A linear fitting method was used to gain the performance parameter K/r_{eff}. For that, Equation (9) was rewritten with the performance parameter $\Delta p_{cap} * K$:

$$\Delta p_{cap} * K * \frac{1}{h} - \rho g K = \mu \varepsilon \frac{dh}{dt} \tag{10}$$

By defining $x = 1/h$ and $y = dh/dt$, the following equation resulted [34]:

$$y = \underbrace{\frac{\Delta p_{cap} * K}{\mu \varepsilon}}_{slope} * x - \underbrace{\frac{\rho g K}{\mu \varepsilon}}_{intercept} \tag{11}$$

For this equation, a linear fitting could be performed with sets of data of x and y from the rate-of-rise measurements. Since the experiment measured the capillary pumping mass instead of the liquid height, the following relation was used [19]:

$$h = \frac{m}{\rho \varepsilon A} \tag{12}$$

This assumed a constant porosity and cross-section of the wick A along the rise direction of the liquid. The parameters x and y were therefore obtained by:

$$x = \frac{1}{h} = \frac{\rho \varepsilon A}{m}; \quad y = \frac{dh}{dt} = \frac{1}{\rho \varepsilon A} \frac{dm}{dt} \qquad (13)$$

The slope of this fitting line then could then be used to determine the wick performance $\Delta p_{cap} * K$ or K/r_{eff}. For the fitting, x and y-values were taken mostly from the intermediate rising period. At the later stage, the equilibrium height was reached and the values for y became very small.

5. Results and Discussion

Several experimental setups were carried out to evaluate the characteristics of the new wick concepts. For the grooved and porous wick structures, the following investigations were performed:

- Microscopy to evaluate process parameters and contour accuracy.
- Helium pycnometry to calculate the porosity.
- Rate-of-rise experiment to calculate the capillary performance.

Additionally, the porous wick structures were examined by mercury intrusion porosimetry to determine the open pore porosity and the wick performance.

5.1. Additive Manufacturing and Geometrical Analysis

In the following, the manufactured samples were presented using optical microscopy images. The geometries were measured, including the porosities of the porous samples.

5.1.1. Grooved Wick Concepts

One wick sample was manufactured for each groove type with each parameter, as shown in Table 2. The polished cut images of the vertically printed trapezoidal grooved wick samples are displayed in Figure 5. Sample 1, which was manufactured with parameter P1, showed the least conformity with the input geometry. Because of unwanted melting, the resulting grooves were narrow. Samples 2 and 3 had well-developed grooves, but the trapezoidal shape was not completely realized. The sample manufactured with parameter P4 showed good geometrical agreement but with a very rough surface due to the low energy density of the laser tracks. While the rough surface might prove beneficial for its function as a wick, the increased porosity of the outer hull could cause leakage problems. Sample 5 displayed the best results regarding groove shape and surface quality. Measurements of the characteristic dimensions, groove depth, and width, deviated less than 5% from the CAD model. In summary, while it was possible to produce acceptable results by choosing a standard parameter, adapting the process parameters of the contour and edge laser tracks positively affected the shape of the microstructure. With a very low power input, rough and partly porous groove structures could be achieved.

In Figure 6, an overview of the four different grooved wick geometries, all printed with parameter P5, is illustrated. The diameter of the artery in the vertically printed arterial wick was represented well, while the entry channel was too wide. As the channel entry, being the liquid-vapor interface, is important for the function of the heat pipe, this area should be improved. The characteristic dimensions measured for the triangular grooves matched well with the input model. Unwanted melted powder partly blocked some of the grooves, which could hinder the liquid flow in the wick and therefore reduce the heat pipe performance. The rectangular and trapezoidal grooves both showed good results regarding geometric accuracy.

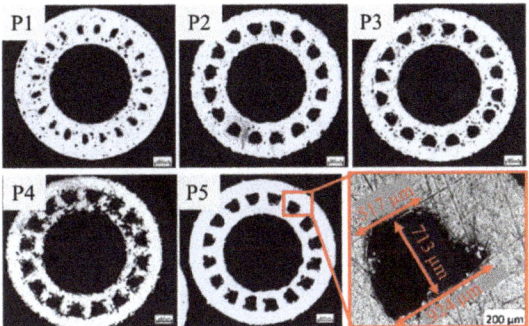

Figure 5. Optical microscopy images of vertically printed trapezoidal grooved wicks with manufacturing parameters P1 to P5.

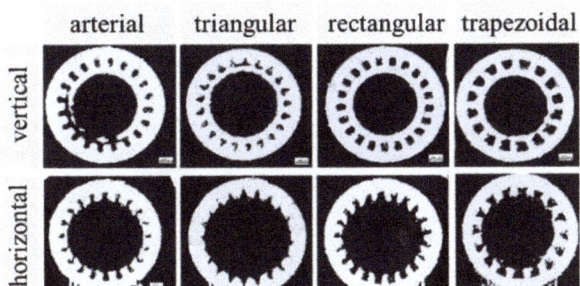

Figure 6. Optical microscopy images of grooved wicks printed vertically (**top row**) and horizontally (**bottom row**) with parameter P5 ($P_{Contour}$ = 300 W; $v_{Contour}$ = 500 mm/s).

For the horizontally printed wick samples, the deviations from the intended shape were higher. The causes are summarized in Figure 7. The grooves at the top of the samples were elongated, while the grooves on the sides were deformed or missing completely. This was an effect of the penetration depth of the laser, which could lead to the unwanted melting of previous layers. Some areas had an overhang angle of over 45° and were only partially supported by the support rings. Increasing the amount of support further would decrease the area of the liquid-vapor interface, which would negatively impact the heat pipe performance. At the same time, it would increase the difficulty of removing the remaining powder from the grooves after the printing process.

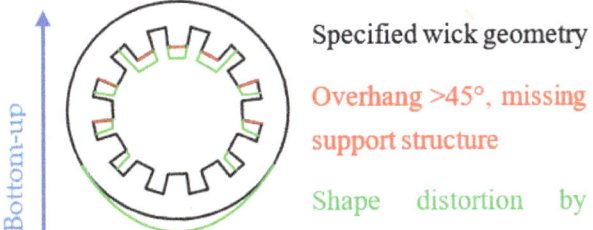

Figure 7. Problems that occur when horizontally printing grooved wick structures.

5.1.2. Porous Wick Concepts

The polished cut images of the produced porous wick samples, printed with parameters P1 to P3, are shown in Table 3 above and depicted in Figure 8. The vertically printed samples showed a separation of the porous wick in the center and the solid outer wall. The dimensions of the wall showed very small deviations compared to the CAD model input of less than 3%. The visible small pores could influence the leak tightness. The wick had a reduced thickness of up to 45% compared to the CAD model, which led to a gap between the inner wall and wick with an average size between 0.21 and 0.27 mm. The reduced volume of the wick had a direct impact on the achievable heat pipe performance. At the same time, the gap could have positively affected the fluid loop by increasing the overall permeability of the wick. Some designs, such as the annular heat pipe, make use of this effect [25]. This influence was less evident for the horizontally printed samples. They also showed a gap in the lower and side areas, while the top part was connected to the outer wall. The porous structure was densified in this area, as the laser melted the subjacent porous layers when it exposed the top of the outer wall.

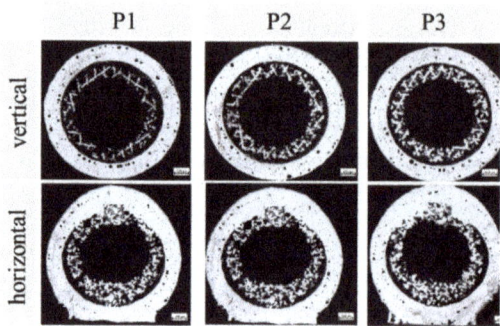

Figure 8. Optical microscopy images of porous wicks printed vertically (**top row**) and horizontally (**bottom row**) with porous process parameters P1 to P3.

A more detailed analysis of the pores was possible with the pore size distribution measured with the mercury intrusion method. The distributions of the horizontally printed porous wick samples are shown in Figure 9. This shows the intruded mercury volume over the pore diameter, normalized on 1 g of the sample. The depicted pores ranged from 3 to 200 µm and were sown in a logarithmic scale.

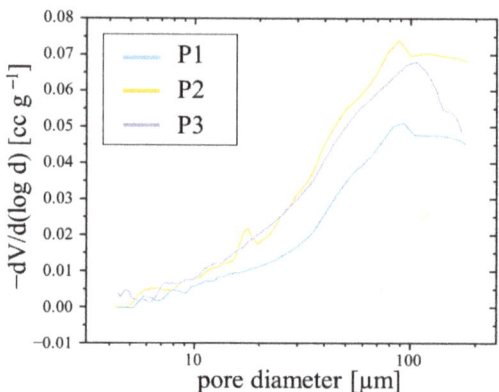

Figure 9. Pore size distribution of the vertically printed porous wick samples P1, P2, and P3.

The graph shows a similar distribution, with the most prevalent pore diameter lying between 88 μm for P2 and 104 μm for P3. The average pore size for all samples was around 45 μm. These pore sizes were relatively small compared to the defined porous SLM structures presented in the literature [8,19]. Small pores benefit the achievable capillary pressure, while in combination with a low porosity also lead to a decrease in wick permeability. The horizontally printed porous samples P1H and P2H showed similar results in the measurements with a most prevalent pore diameter of around 108 μm.

Figure 10 gives an overview of the measured wick porosities: the expected porosity from the preliminary measurements, the theoretical porosity from the helium pycnometry, and the porosity measured from the mercury intrusion. The porosities of the helium pycnometry and mercury intrusion agreed very well and ranged between 20% and 35%, as seen in Figure 10. For samples P1 and P2H, there was a larger deviation, which could stem from pores with a complex intrusion path. These could have been blocked to some extent for the liquid mercury, but accessible for the helium with low viscosity. Both measurements showed a discrepancy of up to 40% from the expected values determined in the preliminary measurements. The reason for this is most likely the limitations of the preliminary analyzation method, which included closed pores when measuring the apparent density. The two setups discussed in this chapter only included open pores, which were reachable by the fluid. Another reason could be the difference in geometry of the porous cube and the wick sample, which might have affected the microstructure and porosity. The trends of the expected porosity and the He-experiment corresponded well with the highest porosity achieved with parameter P1. The effect of vertical and horizontal printing orientation on the porosity was small considering the results from the He-measurement.

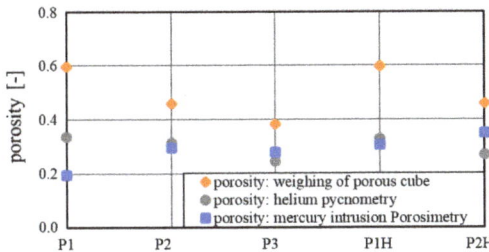

Figure 10. Measured porosities of porous wick concepts by weighting of a porous cube, helium pycnometry, and mercury intrusion porosimetry.

5.1.3. Alternative Wick Concepts

The benefit of the high design freedom of AM was used in the alternative concepts for wick structures. By adding artery and grooves, the porous wick took advantage of their increased permeability and enlarged liquid-vapor-interface, respectively. Figure 11 shows the alternative wick samples printed with parameter P1 (also see Table 3 above). Due to manufacturing restrictions, the results showed no porous structure in these wick samples. Even though the energy density in the wick was reduced in the same way as in the full porous structures, the small dimensions led to a complete melting of the contour tracks. The increased hatch distance could take no effect here, as the contour track, which resolved the groove/artery geometry, dominated. This was also true for the horizontally printed samples. A solution would be to increase the dimensions of the heat pipe or use a metallic powder with a lesser heat conductance. In this way, the geometrical resolution could improve, as the melting pool of the laser tracks becomes smaller.

More promising for an application with the given material and geometrical restrictions is the sloped grooves concept also shown in Figure 11. This showed well-defined grooves both for the vertical and horizontal printing orientation.

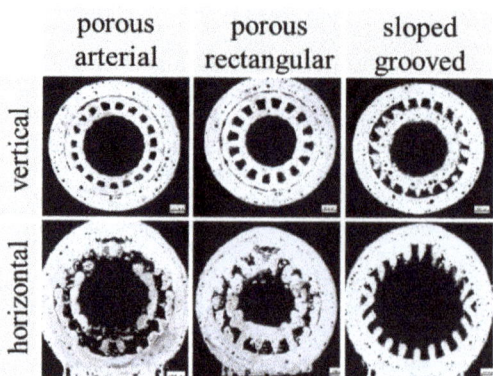

Figure 11. Alternative wick concepts printed vertically (**top row**) and horizontally (**bottom row**).

5.2. Characterization of the Wick Performance

The performance of the grooved wick concepts was characterized using the rate-of-rise method. For the porous concepts, the helium pycnometry and mercury intrusion experiments were additionally used and compared to the rate-of-rise measurements in Section 5.3.

5.2.1. Grooved Wick Concepts

Figure 12 depicts the recorded mass change as a function of time for the grooved wick concepts printed vertically with parameter P5. The triangular grooved wick is not shown, as no capillary pumping could be observed with these samples. The arterial and trapezoidal grooves achieved a similar equilibrium pumping mass, with the arterial groove having a steeper initial rising mass flow. The main rising occurred in the first 5 s, after reaching the equilibrium pumping mass, the mass flow strongly declined.

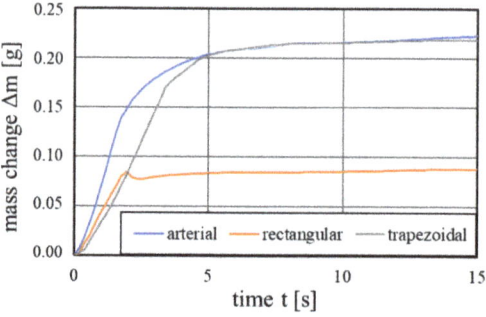

Figure 12. Capillary pumping mass of the vertically printed grooved wicks with the printing parameter 5.

From these mass-time graphs, the following performance parameters were obtained by the previously described linear fitting method from Equations (11)–(13). The arterial and trapezoidal achieved similar performance parameters K/r_{eff} of 1.73 and 1.70 µm, while the rectangular wick achieved 0.28 µm.

The effect of the manufacturing parameter on the grooved wick performance is depicted in Figure 13. While the standard parameters resulted in a limited performance, parameters P4 and P5 showed a vast improvement. While the samples with parameter P5 showed the best match with the input geometry in the polished cut images, the rougher surface of sample P4 appeared to have a positive impact on the capillary pumping. The trapezoidal grooves 4 and 5 achieved a performance parameter of 2.96 and 1.70 µm, respectively. The rectangular grooves also showed the best performance manufactured with these parameters, but were significantly smaller. It appears that the angled channel entry and the higher groove volume of the trapezoidal grooves had a large impact on the wick performance.

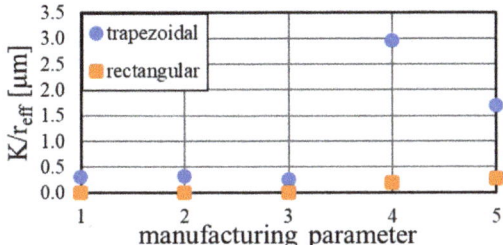

Figure 13. Performance of the trapezoidal and rectangular grooved wick dependent on the manufacturing parameter, calculated by rate-of-rise experiments.

5.2.2. Porous Wick Concepts

Regarding the heat pipe performance, the lower porosity reduced the expected wick permeability. The achieved porosities were in the range of conventionally sintered heat pipes [35] and therefore could have led to a similar performance. The resulting performance parameters are shown in Figure 14. The performance K/r_{eff} of up to 3.22 µm for sample P2 was the same order of magnitude as a comparable study [19], which achieved values between 1.04 and 7.14 µm. The performance of the horizontally printed samples here was around a factor 2 lower than the vertically printed ones, indicating a strong influence of the printing orientation on the permeability of the porous wicks.

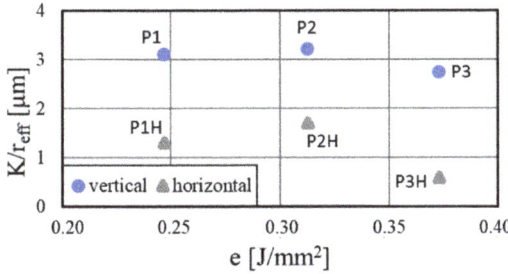

Figure 14. Performance of the porous wicks for different printing parameters and orientations calculated by rate-of-rise experiment.

5.2.3. Alternative Wick Concepts

Finally, the results for the alternative wicks are summarized in Table 4. The porous grooved wick achieved a performance similar to the regular grooved samples, while showing no pumping for the horizontally printed sample. Due to the very small arteries, the porous arterial wick had a very low performance but achieved a better result for the horizontal orientation. As seen in the polished cut image in Figure 11, the wick showed slightly porous parts and a groove-like structure, which could explain the comparably good performance result. The sloped grooves accomplished satisfying values, especially in the first revision. A high groove volume, as is present in these samples, apparently had a positive influence on the resulting performance.

Table 4. Performance of the alternative wick concepts.

K/r_{eff} [µm]	Porous Grooved Wick	Porous Arterial Wick	Sloped Grooves	Sloped Grooves, 1st Rev.
vertical	1.450	0.017	1.971	8.943
horizontal	0.000	2.277	1.530	-

5.3. Comparison of the Measurement Methods

The two investigation methods presented in this paper, the porosimetry by mercury intrusion and the rate-of-rise measurements, both enabled the characterization of the heat pipe performance with the parameter K/r_{eff}. For the mercury intrusion porosimetry, the measured permeability values were divided by the effective pore radius r_{eff}. This radius was gained by $r_{eff} = r_p/cos\theta$, with the measured average pore radius r_p and the contact angle of water-aluminum θ of 45° [25]. As is shown earlier, the rate-of-rise measurements provided K/r_{eff} by the linear fitting. The resulting values as a function of the specific energy density (P1: 0.25 J/mm²; P2: 0.31 J/mm²; P3: 0.37 J/mm²) are compared for the vertically and horizontally printed porous wick samples in Figure 15.

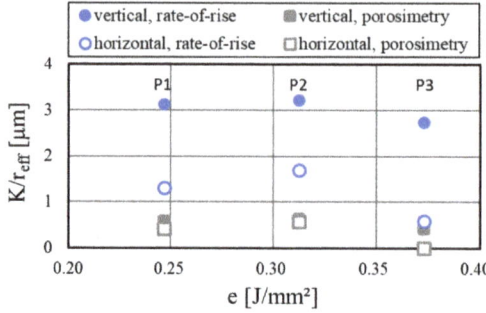

Figure 15. Comparison of the porosimetry and rate-of-rise measurements for the porous wick samples.

A large deviation of absolute values could be observed when comparing the two measurement methods, with the performance parameter gained from the porosimetry being considerably lower. Possible reasons are the differences in the data reduction approaches, which both rely on different assumptions. For the porosimetry, the permeability was modeled with cylindrical pores, while the pore radius was measured directly. In contrast, the rate-of-rise experiment was modeled with a one-dimensional and steady-state laminar flow in the wick and needed more external input values like fluid properties. While the absolute values differed, the trends of both measurement methods agreed very well. With both setups, peak performance at a medium energy density could be observed. The reduction of performance when printing in horizontal orientation could also be resolved with both methods but was more apparent in the rate-of-rise measurements. This was probably due to the influence of the flow direction, which was not considered in the porosimetry measurements. For the characterization of the performance of different wick concepts, both methods were applicable on their own, while the comparability of their absolute results was limited. At the same time, the comparison of the different concepts should not overly rely on the measured K/r_{eff} value. While it is a good indicator of performance for variations of the same concept, the different working principles of the concepts might not be properly be displayed by these measurements. To investigate the wick performance more closely, more complex test setups such as thermal conductance experiments are required.

6. Conclusions

In this work, new concepts for LPBF additively manufactured wick structures were investigated. Various designs were developed which exploited the design freedom of additive manufacturing. The wick structures were determined by geometric parameters, process parameters, or a combination of both. Furthermore, the influences of the printing direction on the wick structures were determined. For all the different test specimens, the characteristic capillary parameters porosity, permeability, effective pore radius, and capillary performance were determined and discussed. Thereby, the rate-of-rise experiment, helium pycnometry, and mercury intrusion porosimetry were applied and the influence on the results was identified. The main conclusions of this work are the following:

- Additive manufacturing with LPBF enables the manufacture of different wick concepts through geometric design, process parameters, or a combination of the two. The grooved wick concepts, especially with trapezoidal and arterial groove geometry, achieve very good wick performances K/r_{eff} of up to 3.0 μm. However, standard manufacturing parameters are not suitable for accurately reproducing the CAD design of filigree structures. It is necessary to develop special process parameters. Furthermore, the porous wick structures achieve a performance K/r_{eff} of 2.7 to 3.2 μm for the vertically printed samples. This demonstrates their great potential, but many closed pores are still suspected. Further adjustments to the process parameters are necessary.
- The printing direction has a significant effect on the printing quality of the wick structures and thus on the capillary performance. Specific process parameters, geometric features such as inner support rings, or adapted geometries such as the sloped grooves, also enable printing in the horizontal direction. Specifically, sloped grooves show an extraordinary performance parameter K/r_{eff} of 8.9 μm, which demonstrates the potential of this concept. This is especially promising when using heat pipes with complex profile shapes, where it is not always possible to manufacture in the optimum orientation.
- Different measurement methods for evaluating the wick performance show large deviations and are therefore only comparable with each other to a limited extent. Whereas measurements with the aid of porosimetry cannot take into account the flow direction of the liquid, the rate-of-rise experiment can, which is important for the comparison of the different print orientations. However, the experiments here showed a good initial comparison between the different concepts, but to enable an accurate evaluation of the wick concepts, thermal conductance and heat transfer limit experiments of filled heat pipes are essential.
- The production of filigree wick structures from metals with high thermal conductivity is challenging. The use of different aluminum alloys enables a higher printing resolution, which provides further optimization potential for the geometrically defined wick structures.

In summary, the presented study showed the great potential of additive manufacturing with LPBF of wick structures, which in the future could allow the heat pipes to be produced in a single manufacturing step. Furthermore, the wick performance, and thus the heat conduction of the heat pipes, can be further increased through adapted and optimized wick designs.

Author Contributions: Conceptualization, K.K.; methodology, K.K., M.B., K.M. and A.P.; investigation, M.B. and P.P.C.H.; writing—original draft preparation, K.K. and M.B.; writing—review and editing, K.K. and K.H.; visualization, K.K. and M.B.; supervision, K.H. All authors have read and agreed to the published version of the manuscript.

Funding: This research received no external funding.

Institutional Review Board Statement: Not applicable.

Informed Consent Statement: Not applicable.

Data Availability Statement: The data and results involved in this study have been presented in detail in the paper.

Conflicts of Interest: The authors declare no conflict of interest.

References

1. Zohuri, B. *Heat Pipe Design and Technology: Modern Applications for Practical Thermal Management*, 2nd ed.; Springer International Publishing: Cham, Switzerland, 2016; ISBN 978-3-319-29840-5.
2. Faghri, A.; Zhang, Y. *Fundamentals of Multiphase Heat Transfer and Flow*; Springer International Publishing: Cham, Switzerland, 2020; ISBN 978-3-030-22136-2.
3. Jouhara, H.; Chauhan, A.; Nannou, T.; Almahmoud, S.; Delpech, B.; Wrobel, L.C. Heat pipe based systems—Advances and applications. *Energy* 2017, *128*, 729–754. [CrossRef]
4. McDonough, J.R. A perspective on the current and future roles of additive manufacturing in process engineering, with an emphasis on heat transfer. *Therm. Sci. Eng. Prog.* 2020, *19*, 100594. [CrossRef]
5. Gibbons, M.J.; Marengo, M.; Persoons, T. A review of heat pipe technology for foldable electronic devices. *Appl. Therm. Eng.* 2021, *194*, 117087. [CrossRef]
6. Esarte, J.; Blanco, J.M.; Bernardini, A.; San-José, J.T. Optimizing the design of a two-phase cooling system loop heat pipe: Wick manufacturing with the 3D selective laser melting printing technique and prototype testing. *Appl. Therm. Eng.* 2017, *111*, 407–419. [CrossRef]
7. Deng, D.; Tang, Y.; Huang, G.; Lu, L.; Yuan, D. Characterization of capillary performance of composite wicks for two-phase heat transfer devices. *Int. J. Heat Mass Transf.* 2013, *56*, 283–293. [CrossRef]
8. Ameli, M.; Agnew, B.; Leung, P.S.; Ng, B.; Sutcliffe, C.J.; Singh, J.; McGlen, R. A novel method for manufacturing sintered aluminium heat pipes (SAHP). *Appl. Therm. Eng.* 2013, *52*, 498–504. [CrossRef]
9. Gupta, M. 3D Printing of Metals. *Metals* 2017, *7*, 403. [CrossRef]
10. Pfaff, A.; Jäcklein, M.; Hoschke, K.; Wickert, M. Designed Materials by Additive Manufacturing—Impact of Exposure Strategies and Parameters on Material Characteristics of AlSi10Mg Processed by Laser Beam Melting. *Metals* 2018, *8*, 491. [CrossRef]
11. Szymanski, P.; Law, R.; McGlen, R.J.; Reay, D.A. Recent Advances in Loop Heat Pipes with Flat Evaporator. *Entropy* 2021, *23*, 1374. [CrossRef]
12. Jafari, D.; Wits, W.W. The utilization of selective laser melting technology on heat transfer devices for thermal energy conversion applications: A review. *Renew. Sustain. Energy Rev.* 2018, *91*, 420–442. [CrossRef]
13. Mullin, N.A.; Galagan, D.V.; Kalyaev, V.Y.; Grol, M.S. Development of a thermal design model for a small spacecraft with integrated heat pipes. In Proceedings of the 71st International Astronautical Congress (IAC)—The CyberSpace Edition, Online, 12–14 October 2020.
14. Robinson, A.J.; Colenbrander, J.; Deaville, T.; Durfee, J.; Kempers, R. A wicked heat pipe fabricated using metal additive manufacturing. *Int. J. Thermofluids* 2021, *12*, 100117. [CrossRef]
15. Kappe, K.; Bihler, M.; Morawietz, K.; Pfaff, A.; Bierdel, M.; Huber, J.; Paul, T.; Hoschke, K. Investigation of additively manufactured structurally integrated heat pipes for CubeSats. In Proceedings of the 72nd International Astronautical Congress (IAC), Dubai, United Arab Emirates, 25–29 October 2021.
16. Szymanski, P.; Mikielewicz, D. Additive Manufacturing as a Solution to Challenges Associated with Heat Pipe Production. *Materials* 2022, *15*, 1609. [CrossRef]
17. Jafari, D.; Wits, W.W.; Geurts, B.J. Phase change heat transfer characteristics of an additively manufactured wick for heat pipe applications. *Appl. Therm. Eng.* 2020, *168*, 114890. [CrossRef]
18. Jafari, D.; Wits, W.W.; Vaneker, T.H.; Demir, A.G.; Previtali, B.; Geurts, B.J.; Gibson, I. Pulsed mode selective laser melting of porous structures: Structural and thermophysical characterization. *Addit. Manuf.* 2020, *35*, 101263. [CrossRef]
19. Jafari, D.; Wits, W.W.; Geurts, B.J. Metal 3D-printed wick structures for heat pipe application: Capillary performance analysis. *Appl. Therm. Eng.* 2018, *143*, 403–414. [CrossRef]
20. Holley, B.; Faghri, A. Permeability and effective pore radius measurements for heat pipe and fuel cell applications. *Appl. Therm. Eng.* 2006, *26*, 448–462. [CrossRef]
21. Chang, C.; Han, Z.; He, X.; Wang, Z.; Ji, Y. 3D printed aluminum flat heat pipes with micro grooves for efficient thermal management of high power LEDs. *Sci. Rep.* 2021, *11*, 8255. [CrossRef]
22. Bierdel, M.; Hoschke, K.; Pfaff, A.; Jäcklein, M.; Schimmerohn, M.; Wickert, M. Multidisciplinary Design Optimization of a Satellite Structure by Additive Manufacturing. In Proceedings of the 68th International Astronautical Congress, Adelaide, Australia, 25–29 September 2017.
23. Bierdel, M.; Hoschke, K.; Pfaff, A.; Schimmerohn, M.; Schäfer, F. Towards flight qualification of an additively manufactured nanosatellite component. In Proceedings of the 69th International Astronautical Congress, Bremen, Germany, 1–5 October 2018.
24. Schimmerohn, M.; Bierdel, M.; Gulde, M.; Sholes, D.; Pfaff, A.; Pielok, M.; Hoschke, K.; Horch, C. Additive Manufactured Structures for the 12U Nanosatellite ERNST. In Proceedings of the 32nd Annual AIAA/USU Conference on Small Satellites, Dublin, Ireland, 4–9 August 2018.
25. Stephan, P. *Wärmerohre: Handbuch Vakuumtechnik*; Springer Fachmedien: Wiesbaden, Germany, 2017.
26. Manakari, V.; Parande, G.; Gupta, M. Selective Laser Melting of Magnesium and Magnesium Alloy Powders: A Review. *Metals* 2017, *7*, 2. [CrossRef]
27. AP Works. *Scalmalloy® Aluminum-Magnesium-Scandium Alloy Data Sheet*; Airbus Apworks GmbH: Taufkirchen, Germany, 2017.
28. Keulen, J. Density of porous solids. *Mat. Constr.* 1973, *6*, 181–183. [CrossRef]

29. Quantachrome Instruments. *Poremaster: Automated Mercury Porosimeters*; Quantachrome Instruments: Boynton Beach, FL, USA, 2018.
30. Washburn, E.W. The Dynamics of Capillary Flow. *Phys. Rev.* **1921**, *17*, 273–283. [CrossRef]
31. Lowell, A.; Shields, V. *Powder Surface Area and Porosity*, 3rd ed.; Springer: Dordrecht, The Netherlands, 1991; ISBN 978-90-481-4005-3.
32. Carniglia, S. Construction of the tortuosity factor from porosimetry. *J. Catal.* **1986**, *102*, 401–418. [CrossRef]
33. Quantachrome Instruments. *UltraPyc: True Volume and Density Analyzer*; Quantachrome Instruments: Boynton Beach, FL, USA, 2019.
34. Deng, D.; Liang, D.; Tang, Y.; Peng, J.; Han, X.; Pan, M. Evaluation of capillary performance of sintered porous wicks for loop heat pipe. *Exp. Therm. Fluid Sci.* **2013**, *50*, 1–9. [CrossRef]
35. Nemec, P. Porous Structures in Heat Pipes. In *Porosity—Process, Technologies and Applications*; Ghrib, T.H., Ed.; InTech: London, UK, 2018; ISBN 978-1-78923-042-0.

Article

Studies on Carbon Materials Produced from Salts with Anions Containing Carbon Atoms for Carbon Paste Electrode

Katarzyna Skrzypczyńska [1], Andrzej Świątkowski [2], Ryszard Diduszko [3] and Lidia Dąbek [4,*]

[1] Łukasiewicz Research Network—Industrial Chemistry Institute, 01-793 Warsaw, Poland; katarzyna.skrzypczynska@ichp.pl
[2] Institute of Chemistry, Military University of Technology, 00-908 Warsaw, Poland; andrzej.swiatkowski@wat.edu.pl
[3] Łukasiewicz Research Network—Institute of Microelectronics and Photonics, 01-919 Warsaw, Poland; ryszard.diduszko@itme.edu.pl
[4] Faculty of Environmental, Geomatic and Energy Engineering, Kielce University of Technology, 25-314 Kielce, Poland
* Correspondence: ldabek@tu.kielce.pl; Tel.: +48-41-34-24-889

Abstract: In the presented work, the properties of carbon materials obtained in the reaction of sodium bicarbonate (C-SB) and ammonium oxalate (C-AO) with magnesium by combustion synthesis were investigated. For the materials obtained in this way, the influence of the type of precursor on their properties was analyzed, including: Degree of crystallinity, porous structure, surface topography, and electrochemical properties. It has been shown that the products obtained in magnesiothermic process were found to contain largely the turbostratic carbon forming a petal-like graphene material. Both materials were used as modifiers of carbon paste electrodes, which were then used to determine the concentration of chlorophenol solutions by voltammetric method. It was shown that the peak current determined from the registered differential pulse voltammograms was mainly influenced by the volume of mesopores and the adsorption capacity of 4-chlorophenol for both obtained carbons.

Keywords: combustion synthesis; salts with carbon in molecule; carbon materials; modified CPEs

1. Introduction

The presence of an increasing amount of pollutants in the environment, including micropollutants, entails the development of research in the field of analytical methods. This method should be more sensitive and accurate, and on the other hand, easy to apply, which will allow their use not only in specialized laboratories, but also by various services working for environmental protection. The standardization of the methods used is also important. Among the whole range of analytical methods (classical, including chromatographic, spectrophotometric), electrochemical methods are of significant importance [1], the measurement possibilities of which are related to the availability of specific/selective electrodes. Carbon paste electrodes (CPEs) are of significant importance in this respect due to their simple structure, easy production, low price, and the possibility of multiple uses in electroanalysis. The usefulness of CPEs in electroanalysis can be significantly improved by using their modification consisting in adding a third component to the basic composition of carbon paste (graphite-paraffin/mineral oil) [2]. Carbon modifiers of CPEs (including graphene, carbon nanotubes) have particular advantages, e.g., good contact of their surface with the analyte, high efficiency of analyte accumulation, electrical conductivity.

Research is constantly being carried out on the relationship between their properties and their effectiveness as modifiers. The properties of these materials (carbon modifiers) and the method of their preparation are of significant importance.

One of the methods used is combustion synthesis (self-propagating high-temperature synthesis) with the use of the carbon-containing precursors leads to their destruction and

the formation of various carbonaceous materials like carbon nanoparticles [3], carbon encapsulates [4], exfoliated graphite [5]. About ten years ago, we reported the use of chlorine-containing organic compounds as precursors in combustion synthesis initiated by sodium azide [6,7]: The carbon materials thereby obtained possessed unique structural and surface properties, and the nature of the chlorine-containing substrate played an important role. The physicochemical properties of these carbons, crystallinity, porosity, adsorption capacity, electrochemical behavior, surface chemistry, varied according to the kind of organic precursor used. Recently, metallothermic reduction of oxalic acid [8], magnesium oxalate [9] or ammonium oxalate (acetate) [10] was utilized with magnesium for preparing carbon materials. The properties of the obtained carbons were characterized in some detail, generally apart from the porous structure. Their uses have also not been studied. Only in one study [10], the adsorption of 4-chlorophenol was investigated. All products of the magnesiothermic precess [8–10] were found to contain largely the turbostratic carbon forming a petal-like graphene material.

Taking all this into account, it could be assumed that these materials could prove to be good modifiers for carbon paste electrodes. It also seemed advisable to use also carbon obtained from inorganic salt, in addition to previously produced only from organic salts.

In this work, carbon material was produced using the described method, based on sodium bicarbonate and ammonium oxalate, and the influence of the type of precursors (anions of the salts used) on the physicochemical properties of the obtained carbons was investigated. At the same time, the usefulness of the so obtained carbon materials as modifiers of carbon paste electrodes (CPE) for the determination of the concentration of organic compounds in an aqueous solution was assessed.

Among the environmental pollutants, 4-chlorophenol was chosen as an example of an organic pollutant from the group of halogenophenol derivatives with a high degree of toxicity, occurring in the environment in a wide range of concentrations [11,12].

2. Materials and Methods

In typical a reaction, powdered magnesium (Mg—of particle diameter below 100 μm) and sodium bicarbonate $NaHCO_3$ or ammonium oxalate $(NH_4)_2C_2O_4$ were triturated dry in ceramic mortar, to obtain homogenic mass. Pressed sample of mass about 10 g was placed in steel crucible, and the combustion process was operated in a steel reactor of volume 275 cm^3, filled with argon under pressure 10 bar. Combustion was carried out thermoelectrically with resistance wire held to the surface of substrate mixture. Results of calorimetric measurements gave values of reaction heats as follows (1,2):

$$3Mg + NaHCO_3 \rightarrow 3MgO + C + NaH \qquad Q = 4692 \text{ J/g} \qquad (1)$$

$$5Mg + (NH_4)_2C_2O_4 \cdot H_2O \rightarrow 5MgO + 2C + 2NH_3 + 2H_2 \qquad Q = 4344 \text{ J/g} \qquad (2)$$

Raw reaction products were cleaned by hours of leaching with concentrated hydrochlorid acid, and next boiling in water. Obtained carbon materials were further referred to as C-SB (sodium bicarbonate) and C-AO (ammonium oxalate).

2.1. Diffraction Analysis

X-ray diffraction (XRD) pattern was measured (D500 Diffractometer, Siemens, Karlsruhe, Germany) using Cu Kα radiation, in the 2Θ range 10–60° for raw reaction products and for cleaned reaction products with a 0.05° step.

2.2. Porosity and Texture

The porosity of the cleaned combustion products was characterized by low-temperature nitrogen adsorption, the relevant isotherms of all samples were measured at 77.4 K on the ASAP 2010 volumetric adsorption analyzer (Micromeritics, Norcross, GA, USA). Before each adsorption measurement, the sample was outgassed under vacuum at 200 °C.

Scanning electron micrograph study was performed with using DSM 942 (Carl Zeiss, Jena, Germany) scanning electron microscope (SEM) for EHT = 2.00 kV.

2.3. Voltammetric Investigations

The electrochemical experiments were carried out using an Autolab potentiostat/galvanostat (model PGSTAT 20, Eco Chemie B.V., Utrecht, The Netherlands) connected to a desktop computer and controlled by a GPES 4.9 software(Eco-Chemie, Utrecht, Netherland. All experiments were carried out in a conventional three-electrode system. The electrode system contained as working electrode carbon paste electrode, a platinum wire as a counter electrode, and a saturated calomel electrode as a reference electrode.

2.4. Preparation of Carbon Paste Electrodes

The bare carbon paste electrode was prepared by mixing 65% of graphite powder (diameter < 20 μm) and 35% of high purity paraffin oil (both components from Sigma-Aldrich, St. Louis, MO, USA) in an agate mortar by hand mixing for about 20 min to get homogenous carbon paste. The paste was packed into the cavity of Teflon electrode body and smoothened on weighing paper.

The modified carbon paste electrodes [2] were prepared by mixing graphite powder and modifier (5 or 10%) with paraffin oil. Homogenization is then achieved by careful mixing using agate pestle and mortar and afterwards rubbed by intensive pressing with the pestle. The mixture was kept at room temperature for two days. The ready-prepared paste was then packed into the hole of the electrode body and the carbon paste was smoothed onto a paper until it had a shiny appearance.

3. Results and Discussion

3.1. Structure and Porosity of Obtained Carbon Materials

XRD spectra recorded for as obtained raw carbon materials C-SB and C-AO as well as cleaned (HCl, H_2O) ones presented in Figures 1 and 2 revealed effects of byproducts removing.

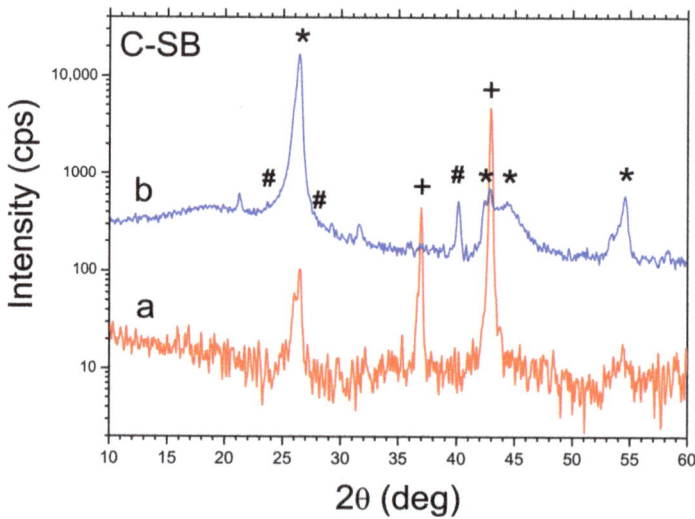

Figure 1. XRD patterns of C-SB (sodium bicarbonate), a—raw reaction product, b—cleaned. Symbols: * graphite, + MgO, # Mg_2Si.

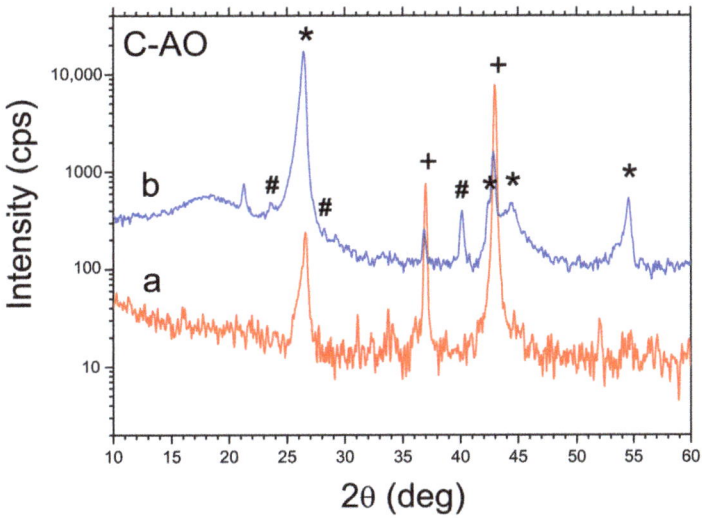

Figure 2. XRD patterns of C-AO (ammonium oxalate), a—raw reaction product, b—cleaned. Symbols: * graphite, + MgO, # Mg$_2$Si.

It has been shown that the cleaning procedure used effectively removes the MgO compound from the structure (in the case of CA-O a trace amount remains, perhaps due to the occlusion phenomenon), while in both cases, a trace amount of Mg$_2$Si appears in the structure. The calculated parameters characterized the structure of both carbon materials (interlayer spacing of crystalline structure d$_{002}$ and number of layers NC) are collected in Table 1. For comparison, in Table 1, both parameter values for graphite G-SA used in CPEs are shown. These data show that the structure of both obtained carbon materials is intermediate between typical for graphite and carbon black [13–16]. The values of the interlayer spacing d$_{002}$ of the crystal structure and the number of layers NC presented in Table 1 show relatively small differences of these parameters for both carbons (C-AO and C-SB). Significant differences between the parameters characterizing the porous structure of both obtained carbons concern mainly the specific surface area and the volume of the mesopores, for C-AO their values are respectively 1.28 and 2.31 times greater than for C–BS.

Table 1. Characteristics of structure and porosity of obtained carbon materials.

Carbon Material	d$_{002}$ (nm)	NC	S$_{BET}$ (m^2/g)	C	V$_{mi}$ (cm^3/g)	V$_{me}$ (cm^3/g)
G-SA	0.3357	>200	4.5	160	0.0008	0.007
C-SB	0.3362	62	39	137	0.005	0.062
C-AO	0.3361	66	50	110	0.005	0.143

The parameters of the porous structure of the obtained carbons C-SB and C-AO, i.e., the specific surface areas (S$_{BET}$) and micropore volumes (V$_{mi}$) as well as mesopore volumes (V$_{me}$) calculated from determined low-temperature nitrogen adsorption isotherms (Figure 3 and for G-SA in [17]) are presented in Table 1. The values of BET surface areas, as well as micro- and mesopores volumes, are less than for typical carbon blacks or graphitized carbon blacks [15] but higher than for graphite. Pore size distribution for both carbon materials, as well as for graphite used in carbon paste electrodes, are shown in Figure 4. The pore size distribution for carbons C-AO and C-SB shows similarity for both carbons and reflects the differences in their mesopore volumes. The major part of the mesopore volume ranges in size from about 23 to 50 nm and further enters the macropore region. The second narrower range of mesopore sizes is 2–17 nm, which, however, has a much smaller pore volume.

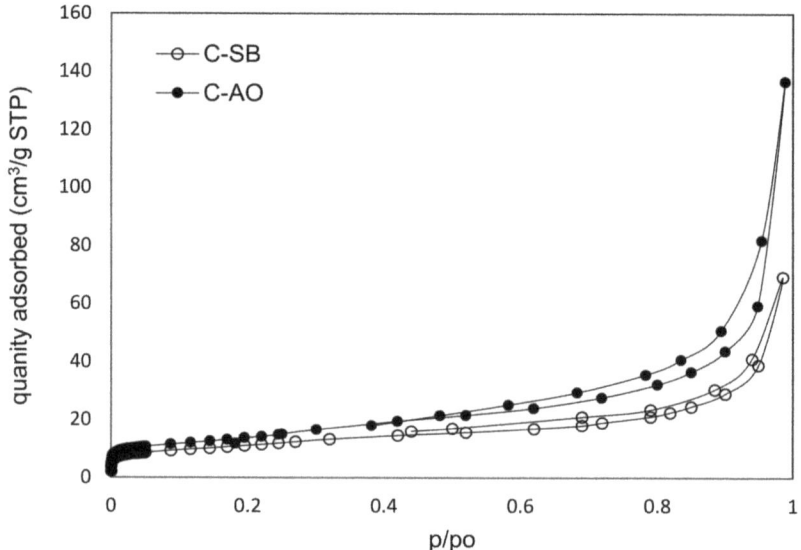

Figure 3. Nitrogen adsorption-desorption isotherms on cleaned carbon materials at 77 K.

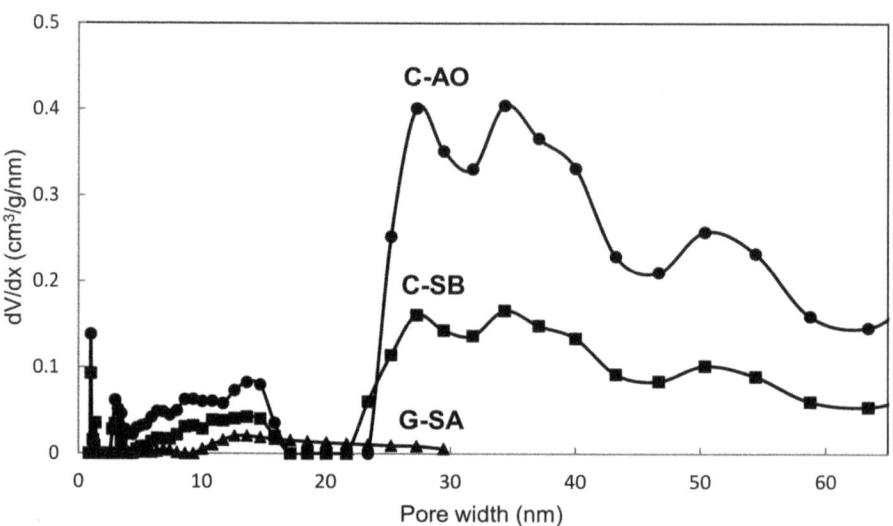

Figure 4. Pore size distribution of obtained cleaned carbon materials as well as graphite used in CPEs.

Surface texture shown in SEM micrographs is given in Figures 5 and 6. The SEM images of both carbon materials are similar and represent a petal-like grapheme material.

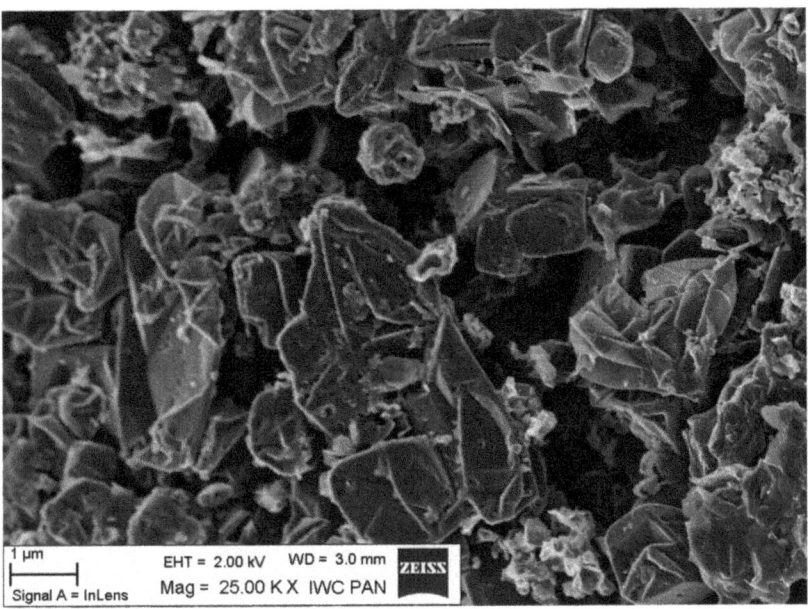

Figure 5. SEM image of carbon material C-SB obtained from NaHCO$_3$.

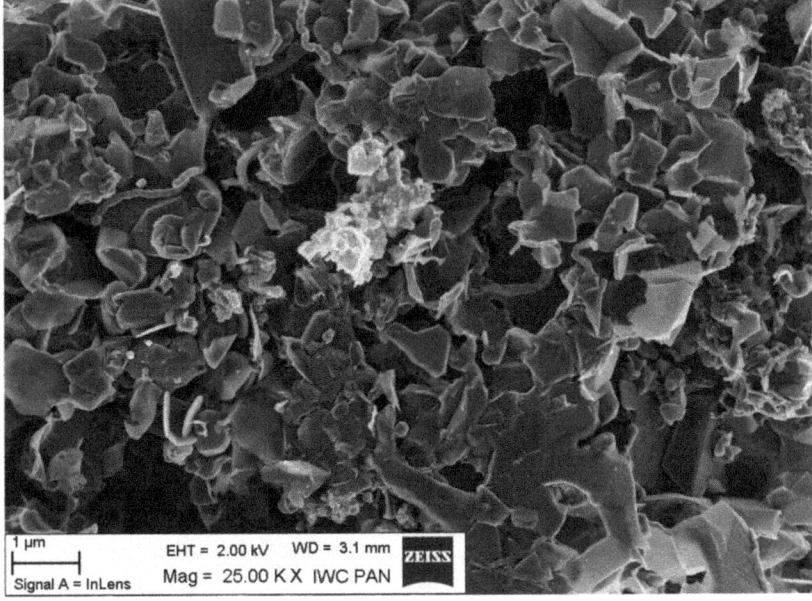

Figure 6. SEM image of carbon material C-AO obtained from $(NH_4)_2C_2O_4$.

3.2. Influence of the CPE Modifier on the Voltammetric Measurements Results

Differential pulse voltammograms (DPV) for the modified CPEs as an example with 0.5 mmol/L 4-CP solutions are shown in Figure 7.

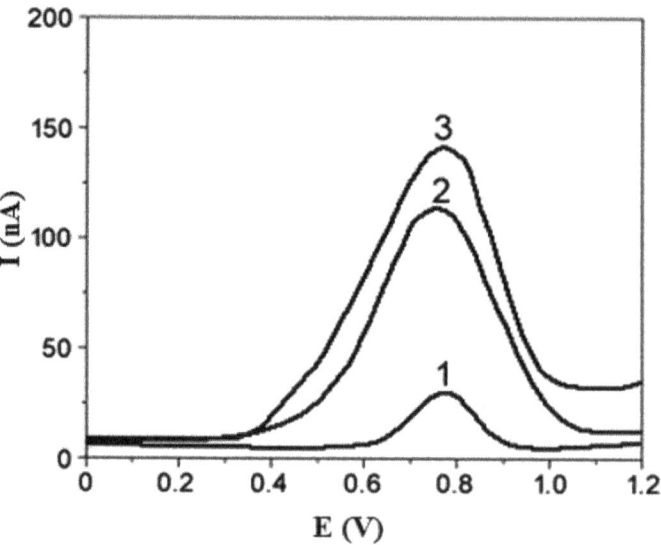

Figure 7. DPV curves registered for 0.5 mmol/dm³ 4-CP solutions in 0.1 mol/dm³ Na₂SO₄ using carbon paste electrodes containing 10 wt.%. of tested materials-modifiers: 1—bare CPE, 2—C-SB, 3—C-AO.

From all the DPV curves (CPEs without as well as with 5% and 10 wt.%. modifiers content) the peak currents I_p and the peak potentials E were determined. All the obtained values are presented in Table 2. The peak potentials reveal slightly different values from 0.78 to 0.79 V for all the used 4-CP concentrations and both modifier's contents or the unmodified CPE. The recorded DPV curves show the dependence of the peak current on the 4-CP solution concentration (Figure 8).

Table 2. Peak currents and potentials determined from DPV curves for carbon paste electrodes modified by adding various quantity of obtained carbon materials.

Modifier Materials	Modifier Content (%)	Concentrations of 4-Chlorophenol Solutions (mmol/L)				Peak Potential (V)
		Peak Current (nA)				
		0.50	0.30	0.20	0.10	
bare CPE	-	24	20	14	9	0.79 ± 0.02
C-SB	5	68	41	29	18	0.79 ± 0.02
	10	103	76	52	31	0.79 ± 0.02
C-AO	5	97	71	50	32	0.78 ± 0.02
	10	152	121	87	57	0.78 ± 0.02

The calibration relationships (peak current versus 4-chlorophenol concentration) were fitted for all CPEs by linear least squares regression analysis. The equations obtained by us, as well as the regression coefficients, are presented in Figure 8. A linear relationship is observed for all CPEs ($R^2 > 0.98$). The I_p exhibit increasing values with an increasing 4-CP concentration for each of the used content of CPE modifier. The peak currents in the case of each modified CPE are about 1.5–2 times higher when the 4-CP concentration is increased two times. The results collected in Table 2 also show the other dependence. The increase of the CPE modifier content gives an increase in the peak current. When the modifier content was enhanced (from 5% to 10 wt.%) one can observe a peak current increase of about 1.7 times. Significant differences in peak current also give the kind of used modifier, obtained carbon material. For each of both contents of modifier and each used 4-CP concentration the C-AO gives higher I_p values than C-SB by about 1.5–1.7 times. The specific surface area of C-AO is about 1.3 times higher than of C-SB. The V_{mi} of both carbons are almost equal but V_{me} is about two times higher for C-AO. It means that the width of pores is important (for both carbon materials in the range of about 25–70 nm).

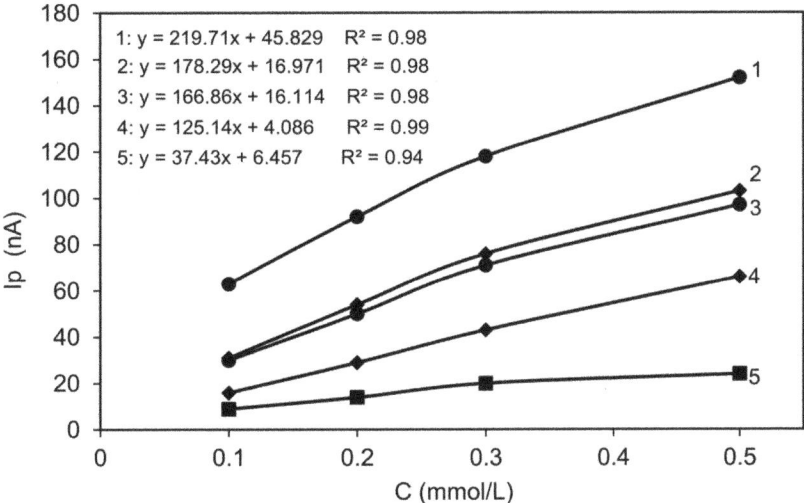

Figure 8. The dependence of peak current in DPV curves on 4-CP concentration for: 1—C-AO 10 wt.%, 2—C-SB 10 wt.%, 3—C-AO 5 wt.%, 4—C-SB, 5 wt.%, 5—bare CPE.

Similar dependence I_p on S_{BET} was previously observed in the case of activated carbon and carbon black materials used as CPE modifiers [16,18]. Increasing the surface area enhanced peak currents for CPE in 4-chlorophenol solutions. For another type of electrode, the powdered carbon electrodes in the case of cyclic voltammetric measurements in 4-CP solutions, such kind of dependence was also observed [15].

Another interesting observation concerns the value of the current I_p measured for both carbon materials produced in this work. In our earlier work [16] for the electrode modified with the addition of Vulcan XC72 carbon black (6%) and for the 0.5 M solution of 4-chlorophenol, the peak current value was 79 nA. However, in this work, for 5% of additives of C-SB and C-AO carbon materials as modifiers, the current values were 68 and 97 nA, respectively. So they were similar to those for the XC72 carbon black used as a modifier. This was despite the fact that this carbon black has a specific surface area approximately 4.6 and 5.9 times greater than that of carbons C-AO and C-SB, respectively. This may be due to the difference in the morphology of the carbon black and our carbon materials.

Vulcan XC72 carbon black has an intermediate structure between amorphous and graphitic, called turbostratic structure. The layers of carbon blacks are parallel to each other but not arranged in order, usually forming concentric inner layers. Carbon black is composed of carbon primary spherical particles of diameters about 30 nm fused together by covalent bonds, thus forming aggregates. Several aggregates can interact to give place to a secondary structure known as agglomerate [19].

Comparing the properties of the above-mentioned carbon black [16] and the carbons obtained in this work, it can be seen that the S_{BET} value is not a proper parameter influencing the measured I_p of CPE. The analysis of other parameters showed that the V_{me} of the carbon black is only 1.38 times greater than the C-AO. Another property of the compared carbon materials, i.e., the adsorption capacity of 4-chlorophenol, also shows a great similarity, e.g., for carbon black and carbon C-AO it is close to 0.5 mmol/g, which was determined, respectively, in the works [10,16]. In recent works on modified carbon paste electrodes [20], carbon nanotubes and reduced graphene oxide have been presented as frequently used modifiers. The products obtained in magnesiothermic process were found to contain largely the turbostratic carbon forming a petal-like graphene material. These materials show promise as modifiers for carbon paste electrodes due to the good contact of their surface with the analyte.

4. Conclusions

The aim of the work was to demonstrate that the proposed type of carbon materials can be successfully used as a modifier of carbon paste electrodes for the determination of chlorophenols concentration by electrochemical method.

Compared to other works on obtaining such specific carbon materials by combustion synthesis, the carbons analyzed in this work were obtained using organic and also inorganic salts containing carbon in their anions, which gives them specific features. It has been shown and confirmed that the products obtained in magnesiothermic process were found to contain largely the turbostratic carbon forming a petal-like graphene material. They were characterized by a relatively small specific surface area and volume of micropores, but a relatively large volume of mesopores. This was shown much better for carbon obtained from ammonium oxalate. The peak current, indicating the effectiveness of the carbon paste electrode modifier, recorded by the DPV method for this carbon as the CPE modifier was comparable to that determined for the carbon black with a specific surface area of 4.5 times greater. This indicates that the surface area and volume of the micropores cannot be the only parameters taken as parameters determining the value of the peak current. The usefulness of carbons produced by the magnesiothermic method as effective CPE modifiers has been used to determine the concentration of 4-chlorophenol.

This encourages the use of these electrodes for the determination of other pollutants as well, which may also contribute to the wider use of electrochemical methods in the analysis of environmental samples.

Author Contributions: K.S.—Investigation, Visualization; A.S.—Conceptualization, Methodology, Writing, Supervision; R.D.— Investigation, Visualization; L.D.—Investigation, Project Administration, Writing—review and editing. All authors have read and agreed to the published version of the manuscript.

Funding: The project was funded by the program of the Minister of Science and Higher Education entitled: "Regional Initiative of Excellence" in 2019–2022, project number 025/RID/2018/19, financing amount PLN 12,000,000.

Institutional Review Board Statement: Not applicable.

Informed Consent Statement: Not applicable.

Data Availability Statement: Data sharing is not applicable to this article.

Acknowledgments: The authors are grateful to Sławomir Dyjak from the Military University of Technology for the preparation (combustion synthesis) of the investigated carbon samples.

Conflicts of Interest: The authors declare no conflict of interest.

References

1. Fifield, F.W.; Haines, P.J. (Eds.) *Environmental Analytical Chemistry*, 2nd ed.; Blackwell-Science: Oxford, MS, USA, 2000; ISBN 978-0-632-053834.
2. Švancara, I.; Kalcher, K.; Walcarius, A.; Vytras, K. *Electroanalysis with Carbon Paste Electrodes*; CRC Press, Taylor & Francis Group: Boca Raton, FL, USA, 2012.
3. Cudziło, S.; Bystrzejewski, M.; Lange, H.; Huczko, A. Spontaneous formation of carbon-based nanostructures by thermolysis-induced carbonization of halocarbons. *Carbon* **2005**, *43*, 1778–1782. [CrossRef]

4. Bystrzejewski, M.; Huczko, A.; Lange, H.; Baranowski, P.; Kaszuwara, W.; Cudziło, S.; Kowalska, E.; Rümmeli, M.H.; Gemming, T. Carbon-encapsulated Magnetic Nanoparticles Spontaneously Formed by Thermolysis Route. *Fuller. Nanotub. Carbon Nanostructures* **2008**, *16*, 217–230. [CrossRef]
5. Cudziło, S.; Szala, M.; Huczko, A.; Bystrzejewski, M. Combustion Reactions of Poly(Carbon Monofluoride), (CF)n, with Different Reductants and Characterization of the Products. *Propellants Explos. Pyrotech.* **2007**, *32*, 149–154. [CrossRef]
6. Cudziło, S.; Huczko, A.; Pakuła, M.; Biniak, S.; Swiatkowski, A.; Szala, M. Surface properties of carbons obtained from hexachlorobenzene and hexachloroethane by combustion synthesis. *Carbon* **2007**, *45*, 103–109. [CrossRef]
7. Cudziło, S.; Bystrzejewski, M.; Huczko, A.; Pakuła, P.; Biniak, S.; Swiatkowski, A.; Szala, M. Physicochemical properties of carbon materials obtained by combustion synthesis of perchlorinated hydrocarbons. *Carbon Sci. Technol.* **2010**, *1*, 131–138.
8. Dyjak, S.; Kiciński, W.; Norek, M.; Huczko, A.; Łabędź, O.; Budner, B.; Polański, M. Hierarchical, nanoporous graphenic carbon materials through an instant, self-sustaining magnesiothermic reduction. *Carbon* **2016**, *96*, 937–946. [CrossRef]
9. Huczko, A.; Kurcz, M.; Dąbrowska, A.; Bystrzejewski, M.; Strachowski, P.; Dyjak, S.; Bhatta, R.; Pokhrel, B.; Kafle, B.P.; Subedi, D. Self-propagating high-temperature fast reduction of magnesium oxalate to novel nanocarbons. *Phys. Status Solidi* **2016**, *253*, 2486–2491. [CrossRef]
10. Huczko, A.; Dabrowska, A.; Fronczak, M.; Bystrzejewski, M.; Subedi, D.P.; Kafle, B.P.; Bhatta, R.; Subedi, P.; Poudel, A. One-Step Combustion Synthesis of Novel Nanocarbons via Magnesiothermic Reduction of Carbon-Containing Oxidants. *Int. J. Self-Propagating High-Temp. Synth.* **2018**, *27*, 72–76. [CrossRef]
11. Czaplicka, M. Sources and transformations of chlorophenols in the natural environment. *Sci. Total. Environ.* **2004**, *322*, 21–39. [CrossRef] [PubMed]
12. Olaniran, A.O.; Igbinosa, E.O. Chlorophenols and other related derivatives of environmental concern: Properties, distribution and microbial degradation processes. *Chemosphere* **2011**, *83*, 1297–1306. [CrossRef] [PubMed]
13. McCreery, R.L. Advanced Carbon Electrode Materials for Molecular Electrochemistry. *Chem. Rev.* **2008**, *108*, 2646–2687. [CrossRef] [PubMed]
14. Donnet, J.-B.; Bansal, R.C.; Wang, M.-J. (Eds.) *Carbon Black: Science and Technology*; Marcel Dekker Inc.: New York, NY, USA, 2003.
15. Biniak, S.; Pakuła, M.; Swiatkowski, A.; Kuśmierek, K.; Trykowski, G. Electro-oxidation of chlorophenols on powdered carbon electrodes of different porosity. *React. Kinet. Mech. Catal.* **2015**, *114*, 369–383. [CrossRef]
16. Kuśmierek, K.; Sankowska, M.; Skrzypczyńska, K.; Swiatkowski, A. The adsorptive properties of powdered carbon materials with a strongly differentiated porosity and their applications in electroanalysis and solid phase microextraction. *J. Colloid Interface Sci.* **2015**, *446*, 91–97. [CrossRef] [PubMed]
17. Krajewski, M.; Świątkowski, A.; Skrzypczyńska, K.; Osawaru, O.; Pawluk, K. Iron nanoparticles and nanowires as modifiers of carbon paste electrodes for the detection of traces of copper, lead and zinc ions in water. *Desalination Water Treat.* **2020**, *208*, 322–329. [CrossRef]
18. Kuśmierek, K.; Świątkowski, A.; Skrzypczyńska, K.; Błażewicz, S.; Hryniewicz, J. The effects of the thermal treatment of activated carbon on the phenols adsorption. *Korean J. Chem. Eng.* **2017**, *34*, 1081–1090. [CrossRef]
19. Singh, M.; Wal, R.L.V. Nanostructure Quantification of Carbon Blacks. *C J. Carbon Res.* **2018**, *5*, 2. [CrossRef]
20. Tajik, S.; Beitollahi, H.; Nejad, F.G.; Safaei, M.; Zhang, K.; Van Le, Q.; Varma, R.S.; Jang, H.W.; Shokouhimehr, M. Developments and applications of nanomaterial-based carbon paste electrodes. *RSC Adv.* **2020**, *10*, 21561–21581. [CrossRef]

Article

Improving the Efficiency of Non-Stationary Climate Control in Buildings with a Non-Constant Stay of People by Using Porous Materials

Alexander Shkarovskiy [1,2,*] and Shirali Mamedov [3]

1. Construction Network and Systems Department of Koszalin University of Technology, 75-453 Koszalin, Poland
2. Department of Heat and Gas Supply and Ventilation, Saint Petersburg State University of Architecture and Civil Engineering, 190005 St. Petersburg, Russia
3. Department of Metal and Timber Constructions, Saint Petersburg State University of Architecture and Civil Engineering, 190005 St. Petersburg, Russia; mamedov_am@bk.ru
* Correspondence: szkarowski@wp.pl; Tel.: +48-607-573-241

Abstract: This article presents the results of experimental research on the non-stationary management of the internal climate of buildings with a non-constant stay of people. During the absence of people, a significant drop in air temperature and corresponding energy conservation in heating is possible. The effectiveness of porous building materials is shown, provided that the appropriate characteristics are selected. Daily fluctuations in the outside temperature are completely extinguished by a layer of foam polystyrene insulation. The absence of channel porosity in the structural material of the wall is a guarantee of the stability of its thermal and humidity regime. This, in turn, prevents the development of mold and mildew.

Keywords: porous materials; channel porosity; internal climate; non-constant stay of people; heating; energy conservation

1. Introduction

The use of porous materials can provide new and unexpected possibilities in problems that are considered to be already solved [1]. However, an important factor is the use of a material with correctly selected characteristics [2]. Our article presents the results of using such materials for non-stationary climate control (NSCC) in non-residential buildings. The term NSCC refers to the programmed control of heating, ventilation, and air conditioning systems in non-residential buildings during the absence of people. In this case, the temperature in the rooms is not kept constant, but changes depending on the presence of people.

Heating costs are as significant for the individual consumer as on a national scale. Residential and public buildings account for 41% of total energy consumption, of which almost 60% are the costs of heating and ventilation [3]. Thus, energy conservation in heating is a natural tendency wherever it does not lead to a decrease in the quality of life.

The most inefficient use of energy is observed in buildings with a non-constant stay of people. This category includes mostly public buildings—schools, universities, office buildings, shopping malls, etc. In such places, it is natural to control the indoor climate actively [4]. During the absence of people, there is no need to comply with the thermal comfort requirements; consequently, a significant decrease in air temperature is possible. This is highly efficient and it is a very important, inexpensive, and easily implemented way to increase the energy efficiency of buildings [5]. In the case of two-stage regulation (Figure 1), before the end of the working day, it is enough to turn off the heating system so that, in some cases, up to 5% of the system capacity is left to prevent pipelines freezing.

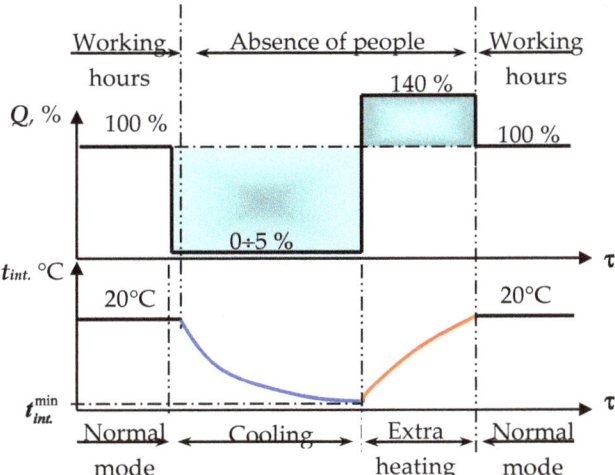

Figure 1. The principle of two-stage climate control (Q—relative power of the heating system; τ—time of day; $t_{int.}$—internal air temperature).

A controlled decrease in air temperature begins in the premises. It is limited only by technical requirements (safety of equipment, plants, works of art in museums, etc.) [6]. If there are no special requirements, then it is important to prevent moisture condensation on the enclosing structures. From this point of view, the minimum permissible temperature is $t_{int}^{min} = 10\ °C$. Before the arrival of people (workers, students, visitors, etc.), the heating is switched on in advance. Provided that it is technically possible, the heating system is switched on with increased power. This will ensure that the temperature rises faster by the start of the working day. The difference indicated by the colored areas in the figure below in the time–power coordinates is a geometric representation of daily energy conservation.

This method of energy conservation has been known for a long time [7]; however, its widespread adoption has been hindered for several reasons. Firstly, there was no theoretical basis and method for active microclimate control calculations. The authors have carried out the necessary theoretical studies in this area. An engineering method for calculating the programmed regulation of a heating system was also developed [8]. The method allows the stages of temperature reduction and subsequent heating to be calculated in relation to the size and design of a building.

Secondly, this method of energy conservation is often confused with room temperature regulation according to heating schedules [9]. The use of heating schedules is a stationary regulation method as it is aimed at long-term maintenance of a certain temperature in rooms. In fact, in this case, the heating system operates with a reduced but constant heat output, depending on the outside and inside temperatures. In the case of active climate control, the heating system is turned off completely, but for a limited time, when there are no people in the premises.

The third reason is the widespread belief that a decrease in indoor temperature can lead to an undesirable change in the temperature distribution in building envelopes. This fact, in turn, can cause subsequent saturation of the enclosing structures with moisture and, as a result, mold and fungal crops can appear.

The last reason should be recognized as the most significant one. The reason for the cases of negative experience of using the NSCC method, according to the authors, is due to a simplified approach to the use of insulation materials. By default, all porous materials are considered to be good thermal insulation materials, though they are not only characterized by high thermal resistance.

Unlike other building materials, they are characterized by a different combination of various thermophysical characteristics. Along with the thermal conductivity coefficient, these include density, thermal diffusivity, porosity, hygroscopicity, and a number of others [10,11]. It is the combination of these characteristics that is important in this case.

In this case, the influence of porosity on moisture permeability is especially important [12,13]. Porous materials have a varied structure, which largely determines the possibility of moisture penetration. There are several types of porosity such as general, closed, and open as well as channel porosity. Channel porosity is especially important from the point of view of moisture permeability since it depends on the closed porosity weakly and is considered to be an independent parameter [14]. Moreover, it is the porosity that determines the gas and moisture permeability of a material [15]. The predominance of channel porosity can be accompanied by moisture migration in building materials, which can lead to undesirable consequences [16,17]. Excess moisture on the wall can even unpredictably affect the operation of heating system devices [18,19]. In this case, the presence of moisture can indeed lead to the appearance of mold and fungal crops. Moisture in microchannels can remain for a long time, even with the next heating of the wall [20]. This may also be facilitated by the formation of hydrates in a limited space of pores [21,22].

In addition, the presence of moisture in microchannels and micropores can have unexpected effects on thermal conductivity. Recent studies have shown that the transfer of the interaction between the solid and the liquid phase contained in the pores to the capillary level and the appearance of hydrogen bonds can dramatically increase the thermal conductivity of a material [23]. It has even been proposed that the consideration of these processes be transferred to the nanoscale, when the effect of aggregation processes occurring between nanoparticles in various aqueous solutions becomes noticeable [24]. In our case, this means a noticeable decrease in the thermal insulation properties of the material and confirms the inadmissibility of the constant presence of moisture in the pores.

Another important result of the latest research is the ability to predict the properties of various materials based on machine learning methods. In this direction, we can note the successful application of the Gaussian process regression (GPR) model to predict delamination in various composites [25]. The use of such methods makes it possible to predict the relationship between predictors of properties of permeable concrete with any of the required properties, including density, porosity, and thermal conductivity [26]. This will allow porous materials with the required properties for subsequent construction to be chosen at the design stage.

In this direction, the authors have undertaken extensive experimental research. The results of one of these experiments are presented in this article. The aim of the research was to prove the safety of the NSCC method when using porous materials with correctly selected characteristics. In this case, the non-stationary control of the internal climate for a limited time, even on weekends, should not lead to a significant change in the temperature distribution in the structures. At the same time, the absence of channel porosity should prevent moisture migration.

2. Design and Realization of Experiments

2.1. Characteristics of the Research Object

For research, a typical building with a non-constant stay of people was chosen —Koszalin University of Technology (Poland). Here, people stay in rooms without outerwear. The type of work is easy mental work. The normalized air temperature in rooms is +20 °C. The city is located in a climatic zone with a design temperature for a heating design of -16 °C. The research was carried out in several adjacent rooms on the top floor of a 7-storey building. The premises, with an area of about 16.5 m^2, has only one external wall, facing southeast. The size of the windows (in the frame) is 1.6×1.6 m, and the frames are metal plastic with double glazing. The ratio of glazing area to floor area is 0.136. In each room, under the window, there is a 1072 W Purmo panel heater equipped with a thermostatic valve. The construction of walls using porous materials is very important for performing research. The four-layer structure includes a bearing layer (aerated concrete blocks), expended polystyrene insulation, and plaster on both sides, as shown in Figure 3a. The thermal and physical parameters of the design (according to the manufacturer's specifications) are given in Table 1. A heat–humidity calculator based on the European standard was used for the calculations [27].

Table 1. Indicators of thermal protection of the external wall structure.

No. of the Layer	Layer Material or Thermal Resistance	Material Density kg/m³	Layer Thickness m	Coefficient of Heat Conductivity W/(m·K)	Heat Resistance m²·K/W
R_{se}	Heat exchange on the external surface	-	-	-	0.040
1	Lime-cement plaster	1850	0.015	0.82	0.018
2	Masonry of aerated concrete blocks on cement mortar	600	0.24	0.16	1.500
3	Foam polystyrene	12	0.10	0.045	2.220
4	Mineral thin-layer plaster	1480	0.02	0.8	0.025
R_{si}	Heat exchange on the inner surface	-	-	-	0.130
	Σ	-	-	-	3.933

Thus, the heat transfer coefficient of the outer wall was $k = 1/3.933 = 0.254$ W/(m²·K). This complies with the European norms for thermal protection of buildings in force from 1 January 2014 (required value: 0.25 W/(m²·K)). However, this indicator is lower than the norms that came into practice on 1 January 2017 (0.23 W/(m²·K)) and those introduced from 1 January 2021 (0.20 W/(m²·K)). An important factor is the manufacturer's guarantee of the absence of through-channel porosity of aerated concrete blocks and expanded polystyrene. In addition, the alkaline reaction of aerated concrete is an additional property that prevents the development of fungal crops.

2.2. Research Program

It was necessary to investigate the efficiency of porous materials in two-stage non-stationary climate control on working days (Figure 1). It was important to prove the safety of the NSCC method for enclosing structures. It was also planned to determine the optimum start time of heating so that the required temperature in the premises was reached by 8 a.m. The following experimental plan was developed when the heating system was turned off one hour before the people left (8 p.m.), and the temperature was continuously recorded from 7 p.m. to 8 a.m [28]. The measurement interval was 10 min, and the temperatures of indoor and outdoor air, temperatures on both surfaces of the wall, and at selected points inside its structure were measured.

2.3. Experimental Unit and Research Technique

The experimental complex was based on the AVT5330 multipoint electronic temperature recorder (Figure 2a) with software for operation in Windows. Automatic measurement is possible at any interval, starting from 2 s.

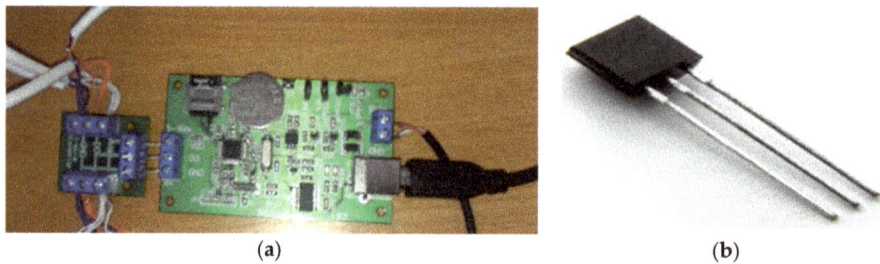

Figure 2. Measurement complex: (**a**) AVT5330 recorder; (**b**) DS18B20 sensor.

The recorder provides the connection of 8 DS18B20 temperature sensors (Figure 2b). The connection is made with a 2-m cable. For protection from external influences, the sensor is insulated with a heat-shrinkable sheath. The sensors were pre-calibrated in a certified laboratory.

In Figure 3, the layout of the sensors is shown. Sensors 2, 3, 5, and 8 were placed inside the wall in channels 8 mm in diameter at different depths. At the same time, sensor 3 recorded the temperature at the boundary of the carrier layer and the insulation (Figure 3a). Sensors 7 and 1 were fixed directly on the inner and outer surfaces of the wall, sensor 4 recorded the air temperature in the room, and sensor 6—the temperature of the outside air at a distance of about 0.5 m from sensor 1.

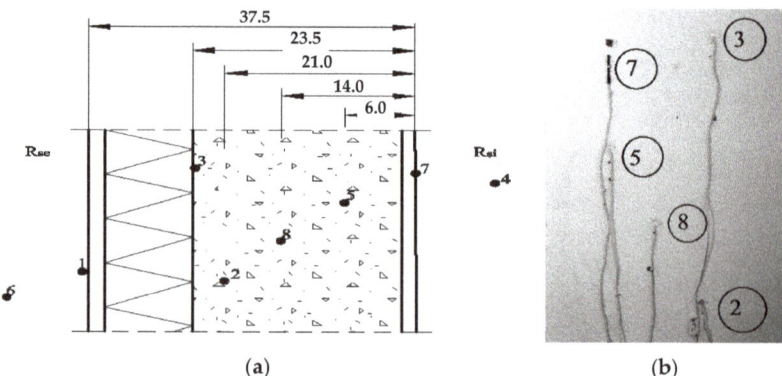

Figure 3. Placement of temperature sensors: (**a**) along the thickness of the enclosing structure (centimeters); (**b**) in the plane of the wall.

The sensors were placed at a sufficient distance from each other (Figure 3b) in order to avoid changes in the measurement results. Channels were drilled from inside the room, and sensors 1 and 6 were passed through the window frame. The volume of the drilled channels was negligible in comparison with the volume of the wall covered by the experiment and could not affect the heat transfer process.

Simulation of the operating modes of the heating system was carried out using Danfoss thermostatic valves. Setting the valve to the minimum position led to the shutdown of the heating device. Setting the valve to the maximum position provided heating with power of up to 150%.

3. Research Results and Discussion

Preliminarily, using our own methodology [29], calculations of temperature changes in the premises were performed, assuming that the outside temperature was −7 °C and there was a possibility of increasing the power of the heating system during heating up to 155%, as shown in Figure 4.

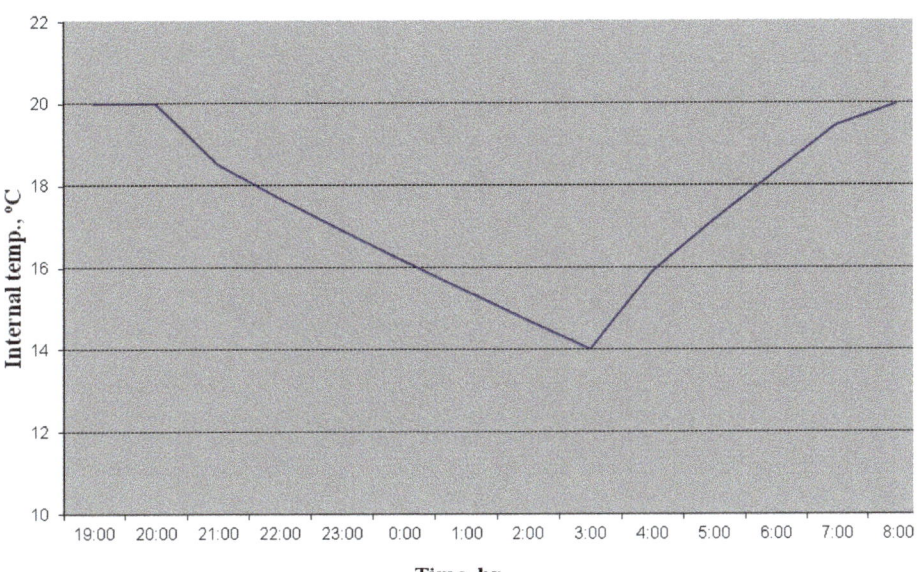

Figure 4. Calculated graph of non-stationary climate control on a working day at an outside temperature $t_{ext.} = -7\ °C$.

After that, experimental studies began according to the program mentioned above.

As an example, the measurement results are given for a typical day with a drop in outdoor temperature to $-2.0\ °C$. Figure 5 shows the temperature change recorded by each sensor during the period when the heating system was turned off. Figure 6 shows the temperature profiles for the selected moments in the same period.

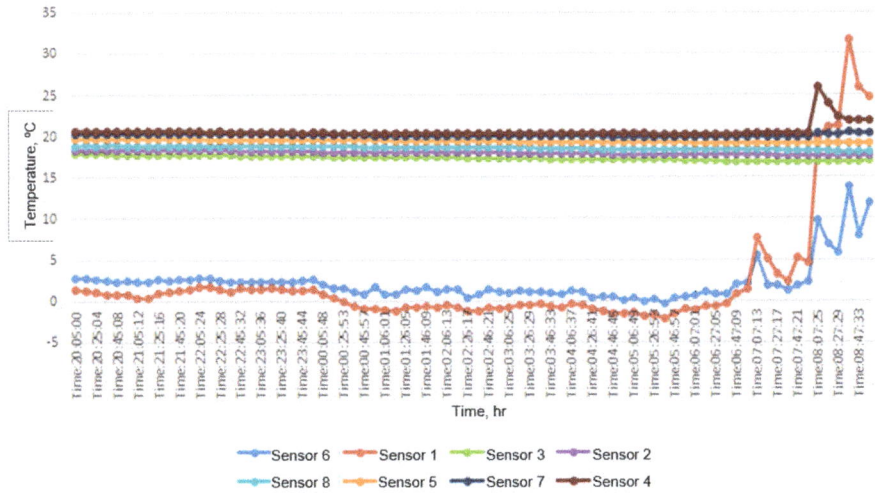

Figure 5. Change in sensor data for a typical measurement day.

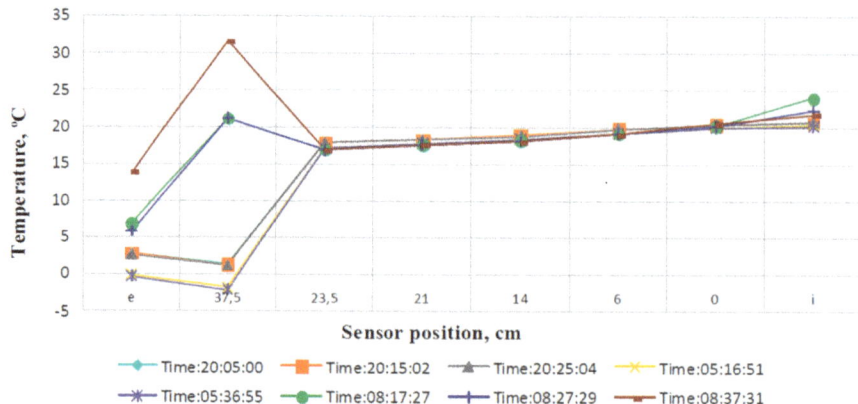

Figure 6. Temperature profiles for the selected measurement times.

The temperature in the room, on the inner surface of the wall, and in its bearing layer did not undergo significant changes, despite the complete shutdown of the heating system and noticeable fluctuations in the outside temperature. Outside temperature fluctuations at night were almost completely extinguished by the layer of foam polystyrene. This is precisely the technological task of thermal insulation. An additional thermal buffer is the carrier layer made of aerated concrete blocks.

Thus, the obtained data confirm the safety of the method of active climate control for enclosing structures. The impossibility of significant penetration of a zone with a temperature below the dew point (about 6 °C) deep into the wall structure has been proved. This makes it impossible to moisten the carrier layer and prevents the further development of unfavorable phenomena (mold, fungal cultures). Thus, the effectiveness of the use of materials that do not have channel porosity was also proved.

Fluctuations in the temperature of the internal air turned out to be much fewer than the calculated ones (Figure 4). This can be explained by the fact that the calculation method considers the internal volume of the building as empty space. In reality, interior walls, ceilings, furniture, and equipment are massive heat accumulators. This softens the temperature drop after the heating system is turned off. An additional factor is the use of porous materials in the construction of the wall.

The rise in the outside temperature in the morning can be explained by the influence of solar radiation. In addition, the extreme changes in the internal temperature after turning on the heating are caused by the proximity of the heater.

Since the temperature drop in these rooms was very small, there was no need to start the heating phase ahead of time, as it was supposed by theory (see Figures 1 and 2). The heating system was switched on at the beginning of the working day. The thermostatic valves were immediately set to the calculated position.

The studies were carried out every winter from 2015 to 2019, when the building structure was additionally insulated. Additional thermal protection only enhanced the achieved effect and the conclusions drawn during the research. The goal of the work was recognized as being achieved and the experiments were terminated.

4. Conclusions

1. Studies have confirmed the ease of use and high economic effect of the method of saving energy through non-stationary control of the internal climate in buildings with a non-constant occupancy of people.
2. It has been proven that despite the complete shutdown of the heating system, there were no noticeable temperature fluctuations in the room, on the wall surface, and inside the bearing layer of the enclosing structures. Night-time fluctuations in the

outdoor temperature were completely extinguished by the insulation layer, which is its technological purpose.
3. The method does not cause moisture migration and permanent moisturizing of porous building and insulation materials. As a result, there is no danger of a significant change in the thermal insulation properties of materials and undesirable development of mold and fungal crops.
4. An additional guarantee of the absence of adverse side effects is the use of building materials that do not have channel porosity. At the same time, modern modeling methods allow us to predict and correctly apply materials with precisely selected properties already at the design stage.
5. The research results indicate a significantly greater effect of the NSCC method in comparison with the theoretical change in temperature. This, firstly, proves the effectiveness of the use of porous materials in the construction of walls. On the other hand, it reveals that it is necessary to make adjustments to the calculation methodology, taking into account the internal structure and equipment of the building.

Author Contributions: A.S.: Conceptualization, investigation, data analysis, writing—original draft; S.M.: Investigation, data analysis, writing—editing. All authors have read and agreed to the published version of the manuscript.

Funding: This research received no external funding.

Institutional Review Board Statement: Not applicable.

Informed Consent Statement: Not applicable.

Data Availability Statement: The data presented in this study is available at the request of the respective author.

Conflicts of Interest: The authors declare no conflict of interest.

References

1. Koshlak, H.; Kaczan, A. The Investigation of Thermophysical Characteristics of Porous Insulation Materials Based on Burshtyn TPP Ash. *Rocz. Ochr. Środowiska* **2020**, *22*, 537–548.
2. Lee, D.J.; Lai, J.Y.; Mujumdar, A.S. Moisture Distribution and Dewatering Efficiency for Wet Materials. *Dry. Technol.* **2006**, *24*, 1201–1208. [CrossRef]
3. *Directive 2010/31/EU of the European Parliament and of the Council of 19 May 2010 on the Energy Performance of Buildings*; Publications Office of the European Union: Luxembourg, 2010.
4. Costanzo, G.; Iacovella, S.; Ruelens, F.; Leurs, T.; Claessens, B. Experimental analysis of data-driven control for a building heating system. *Sustain. Energy Grids Netw.* **2016**, *6*, 81–90. [CrossRef]
5. a, A. a. *a a*; Palmarium Academic Publishing: Saarbrücken, Germany, 2012.
6. Szkarowski. *Ciepłownictwo*; Wydanie III uaktualnione; PWN: Warszawa, Poland, 2019.
7. Shaikh, P.H.; Nor, N.B.M.; Nallagownden, P.; Elamvazuthi, I.; Ibrahim, T. Intelligent multi-objective control and management for smart energy efficient buildings. *Int. J. Electr. Power Energy Syst.* **2016**, *74*, 403–409. [CrossRef]
8. Dyczkowska, M.; Szkarowski, A. Metoda energooszczędnego sterowania pracą instalacji grzewczych w budynkach o podwyższonej izolacyjności cieplnej—porównanie modelu matematycznego z wynikami badań. *Rocz. Ochr. Środowiska* **2009**, *11*, 583–594.
9. Schaub, M.; Kriegel, M.; Brandt, S. Analytical prediction of heat transfer by unsteady natural convection at vertical flat plates in air. *Int. J. Heat Mass Transf.* **2019**, *144*. [CrossRef]
10. Shelby, J.E. *Introduction to Glass Science and Technology*, 2nd ed.; Royal Society of Chemistry: Cambridge, UK, 2005.
11. Nimmo, J.R. Porosity and Pore Size Distribution. In *Encyclopedia of Soils in the Environment*; Hillel, D., Ed.; Elsevier: London, UK, 2004; Volume 3, pp. 295–303.
12. Tarasov, V.E. Heat transfer in fractal materials. *Int. J. Heat Mass Transf.* **2016**, *93*, 427–430. [CrossRef]
13. Cherki, A.-B.; Remy, B.; Khabbazi, A.; Jannot, Y.; Baillis, D. Experimental thermal properties characterization of insulating cork–gypsum composite. *Constr. Build. Mater.* **2014**, *54*, 202–209. [CrossRef]
14. Eom, J.-H.; Kim, Y.-W.; Raju, S. Processing and properties of macroporous silicon carbide ceramics: A review. *J. Asian Ceram. Soc.* **2013**, *1*, 220–242. [CrossRef]
15. Raoof, A.; Nick, H.; Hassanizadeh, S.; Spiers, C. PoreFlow: A complex pore-network model for simulation of reactive transport in variably saturated porous media. *Comput. Geosci.* **2013**, *61*, 160–174. [CrossRef]

16. Pavlenko, A.; Piotrowski, J.Z. Mathematical Model of the Drying Process of Wet Materials. *Rocz. Ochr. Środowiska* **2020**, *22*, 347–358.
17. Wang, Y.; Ma, C.; Liu, Y.; Wang, D.; Liu, J. Effect of moisture migration and phase change on effective thermal conductivity of porous building materials. *Int. J. Heat Mass Transf.* **2018**, *125*, 330–342. [CrossRef]
18. Chen, H.-T.; Lin, Y.-S.; Chen, P.-C.; Chang, J.-R. Numerical and experimental study of natural convection heat transfer characteristics for vertical plate fin and tube heat exchangers with various tube diameters. *Int. J. Heat Mass Transf.* **2016**, *100*, 320–331. [CrossRef]
19. Orłowska, M. Experimental Research of Temperature Distribution on the Surface of the Front Plate, of a Flat Plate Heat Exchanger. *Rocz. Ochr. Środowiska* **2020**, *22*, 256–264.
20. Lin, T.-Y.; Kandlikar, S.G. A Theoretical Model for Axial Heat Conduction Effects During Single-Phase Flow in Microchannels. *J. Heat Transf.* **2011**, *134*. [CrossRef]
21. Okutani, K.; Kuwabara, Y.; Mori, Y.H. Surfactant effects on hydrate formation in an unstirred gas/liquid system: An experimental study using methane and sodium alkyl sulfates. *Chem. Eng. Sci.* **2008**, *63*, 183–194. [CrossRef]
22. Yiotis, A.G.; Tsimpanogiannis, I.N.; Stubos, A.K.; Yortsos, Y.C. Pore-network study of the characteristic periods in the drying of porous materials. *J. Colloid Interface Sci.* **2006**, *297*, 738–748. [CrossRef]
23. Christensen, G.; Lou, D.; Hong, H.; Peterson, G. Improved Thermal Conductivity of Fluids and Composites Using Boron Nitride (BN) Nanoparticles through Hydrogen Bonding. *Thermochim. Acta* **2021**, *700*. [CrossRef]
24. Younes, H.; Hong, H.; Peterson, G.P. A Novel Approach to Fabricate Carbon Nanomaterials–Nanoparticle Solids through Aqueous Solutions and Their Applications. *Nanomanuf. Metrol.* **2021**, 1–11. [CrossRef]
25. Zhang, Y.; Xu, X. Predicting the delamination factor in carbon fibre reinforced plastic composites during drilling through the Gaussian process regression. *J. Compos. Mater.* **2020**. [CrossRef]
26. Zhang, Y.; Xu, X. Predicting Multiple Properties of Pervious Concrete through the Gaussian Process Regression. *Adv. Civ. Eng. Mater.* **2021**, *10*, 56–73. [CrossRef]
27. PN-EN ISO 6946:2004. Building Components and Building Elements. Thermal Resistance and the Heat Transfer Coefficient. Calculation Method (Komponenty Budowlane i Elementy Budynku). Opór Cieplny i Współczynnik Przenikania Ciepła. Metoda Obliczania. 2004. Available online: http://www.vinci-facilities.pl/wp-content/uploads/2018/01/PN-EN-12831.pdf (accessed on 28 April 2021).
28. Szkarowski, A.; Gawin, R. Improving Energy Efficiency of Public Buildings. *Energy Policy J.* **2016**, *19*, 87–97.
29. Shkarovskiy, A. *Efficiency of Energy Saving Investments (Efektywność Inwestycji Energooszczędnych)*; GlobeEdit: Mauritius, Africa, 2018.

Article

Numerical Network Modeling of Heat and Moisture Transfer through Capillary-Porous Building Materials

Borys Basok [1], Borys Davydenko [1] and Anatoliy M. Pavlenko [2,*]

1. Head of the Department of Thermophysical Basics of Energy-Saving Technologies, Institute of Engineering Thermophysics National Academy of Sciences of Ukraine, 03057 Kyiv, Ukraine; basok@ittf.kiev.ua (B.B.); bdavydenko@ukr.net (B.D.)
2. Department of Building Physics and Renewable Energy, Kielce University of Technology, al. Tysiąclecia Państwa Polskiego 7, 25-314 Kielce, Poland
* Correspondence: apavlenko@tu.kielce.pl; Tel.: +48-883-741-291

Abstract: The article presents the modeling of the dynamics of the vapor-gas mixture and heat and mass transfer (sorption-desorption) in the capillary structure of the porous medium. This approach is underpinned by the fact that the porous structure is represented by a system of linear microchannels oriented along the axes of a three-dimensional coordinate system. The equivalent diameter of these channels corresponds to the average pore diameter, and the ratio of the total pore volume to the volume of the entire porous material corresponds to its porosity. The entire channel area is modeled by a set of cubic elements with a certain humidity, moisture content, pressure and temperature. A simulation is carried out taking into account the difference in temperatures of each of the phases: solid, liquid and gas.

Keywords: porous medium; heat transfer; mass transfer; mathematical modeling; numerical research methods

1. Introduction

Most of the materials used in construction have a capillary-porous structure. The thermal insulation properties of these materials depend on the condition parameters: temperature, pressure, humidity and moisture content. Predicting the heat loss levels from the premises to the surrounding space through enclosing structures depends on the accuracy and reliability of heat and mass transfer through the capillary-porous media.

Many computational schemes use models based on the phenomenological theory of mass and heat transfer [1–3], whereby a real porous structure is replaced by a homogeneous continuous medium. The transfer processes for this continuous medium are expressed by mass and energy conservation equations, where volume-averaged physical values and effective transfer coefficients are used [4–7].

This approach is quite justified, as the shape of pores, their quantity and distribution in the material volume are random parameters, if we do not mean formed cracks in pore connecting interpore space or channel porosity. The shape of such pores has a pronounced configuration and size. It is the channel porosity (cracks, as shown in Figure 1) that can significantly change thermophysical properties of the material. Naturally, in this case, averaging of physical values over material volume results in errors in the calculations of heat and mass transfer parameters.

In some cases, the use of this approach to solving problems of heat and mass transfer results in uncertain individual values of transfer equations. In particular, it refers to source terms, included with different signs in liquid and vaporous moisture mass conservation, and expressing the moisture transition rate from one phase to another, during liquid evaporation or condensation inside the material.

As it is difficult to determine this value, both mass conservation equations are usually summed up. The resulting mass transfer equation no longer contains the specified value.

However, in this case, the resulting equation describes the transfer of a certain total moisture content, including both liquid and vapor phases. In this instance, the moisture evaporation or condensation rate inside the material remains in the energy equation. Many researchers use this technique. But at the same time the physics of the effects of evaporation (condensation) remain undisclosed. We, however, avoided the indicated method and directly considered the effects of the phase transition-evaporation or condensation. This is the main idea of the article.

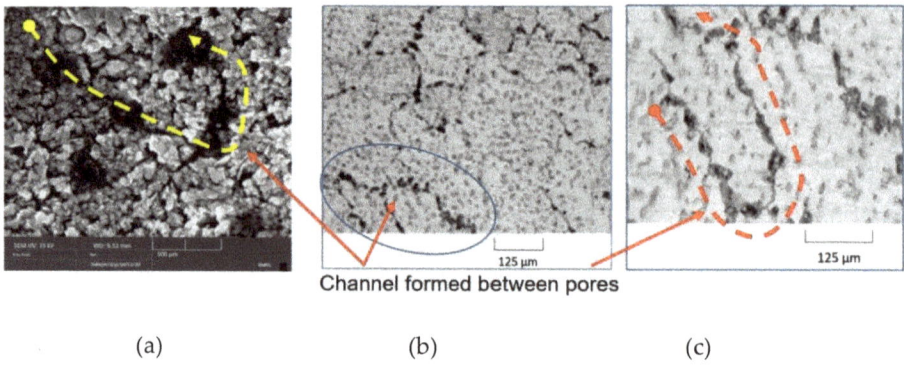

Channel formed between pores

(a)　　　　　　　　　　(b)　　　　　　　　　　(c)

Figure 1. Structure of materials: (**a**) granular concrete; (**b**), (**c**) cellular concrete.

For example, this approach is applied in the work [3], where the authors propose a one-dimensional model, consisting of energy, dry air and total humidity equations.

In the work [5], they propose a mathematical model for the drying of wet building materials, taking into account the presence of water and vapor. Pressure and temperature are taken as variables. The authors consider simultaneous capillary water transfer and vapor diffusion in two-dimensional areas. The effect of dry air movement was not considered in these models. In the work [8], a mathematical model is represented by equations of moisture and heat, transferred through a silica brick; these parameters were taken as independent variables. In the work [9] the same approach is proposed, but moisture and heat are transferred through a complex anisotropic material structure. In the presented works, the models take into account three basic phenomena: vapor diffusion, capillary suction in a porous medium and advective transfer of moist air through thin channels. A similar calculation scheme for moisture transfer in brick is presented in [10] and it is based on the same control potentials.

An expanded mathematical model of heat and mass transfer in the homogeneous porous building materials is presented in [11–13]. It includes four basic transfer equations: water vapor, dry air, liquid moisture and energy. Dry air and water vapor densities, as well as a volume fraction of liquid moisture and temperature, are used as independent variables. The analyzed building material, namely brick, is considered as a porous material. A solid phase is the material from which the brick is made; water and moist air are present in its pores. The amount of water in the building material pores changes as a result of the transfer caused by capillary pressure gradient, as well as evaporation and condensation processes, while the amount of vapor also changes as a result of diffusion and phase transition processes. In the presented models, phase heat equilibrium is assumed, therefore a unified equation of energy transfer is considered. It also assumes averaging the parameters within material volume.

Another approach, used to describe heat and mass transfer processes in capillary-porous materials, is associated with a model of the evaporation zone deepening [14–16]. According to this model, there are dry and moist zones in a wet material. In a dry zone, moisture is present only in a gaseous form (as vapor), and in the moist zone, all pores are occupied by liquid moisture. The liquid evaporates only at the interface of these zones, it is deepening towards the wet moist zone. It is assumed that the heat is supplied to the evaporation boundary by applying thermal conductivity of the material dry layer and spent on moisture evaporation.

The mathematical formulation of this process is based on a Stefan-type problem [17]. Similar models are proposed in the works [14,15]; however, they neither consider radiation heat transfer on the dried surface, nor analyze the step size sensitivity or computational grid density. Currently, the heat and mass transfer models, based on the capillary-porous structure, represented as the so-called pore network, are used [18–23]. According to this model, a real microstructure of the porous material is replaced by a system of interconnected and intersecting channels with a known arrangement and geometric dimensions. Results of the mass transfer study, using this approach, are presented in [24–30].

Figure 2 shows the most common network models, where pores are represented by lines.

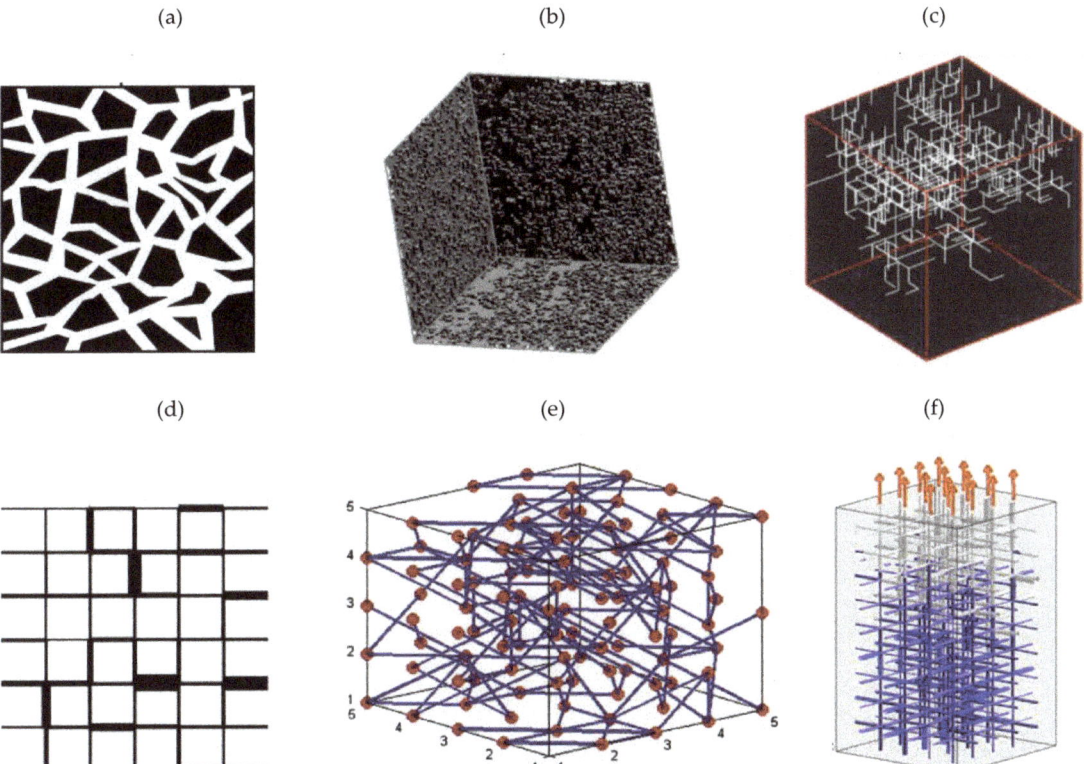

Figure 2. Network models: (**a**)–[4]; (**b**)–[5]; (**c**)–[6]; (**d**)–[7]; (**e**)–[8]; (**f**)–[9].

In these works, several numerical approaches are proposed for modeling the transfer of heat, mass and momentum during porous material dehydration. These approaches are characterized by spatial scale and physical processes to be reflected in the models. These models consider the material as a continuum divided into microvolumes. It is assumed that in these microvolumes (MV) individual phases are superimposed on each other, meaning that they cannot be analyzed separately. Therefore, MV should be large enough, for example larger than the pore size, in order to provide averaging of material properties within the MV. On the other hand, MV should also be small enough to prevent changes in the studied parameters within these volumes (e.g., temperature), resulting from macroscopic gradients and associated nonequilibrium conditions at this microscale level. Transfer inside the material is modeled by averaged material properties, obtained either experimentally, or by numerical calculation. Thus, complex pathways and microscale transfer processes are included in a concentrated way in the material properties and transfer equations, instead of explicitly taking them into account by modeling. A typical example is the use of the Darcy's law combined with fluid permeability, i.e., a macroscopic material property, in order to describe the fluid transfer inside a porous material at the continuum level, inherently including complex transfer phenomena at the microscale level. These material properties are often a complex function of temperature and moisture.

In the works [22,23], it is shown that the model of a porous medium drying zone is the result of generalization of many phenomenological observations and experimental studies, and describes liquid phase distribution during drying of porous media. But they fail to explain the internal mechanism of the "evaporation zone" phenomenon. Namely, which of the drying factors affects liquid phase distribution during drying of porous media? Therefore, in these works, the pore network models are proposed, which are applicable for the slow isothermal drying of porous media.

In the works [24,25], associated heat and mass flows in the voids of complex geometry are considered. The conventional drying models, presented in the above works, are based on the assumption that a porous medium is a fictitious continuum, for which heat and mass balances are derived either by homogenization or by volume averaging. The pore network models are mainly developed because it is impossible to study transport phenomena at the pore level. Therefore, the exact description of a transfer in a porous medium is greatly simplified to the description of individual phases, i.e., gas and liquid.

In the works [25–27,30–33], the unsaturated moisture transfer processes in hygroscopic capillary-porous materials are simulated, demonstrating a wide pore size distribution. The pores are seen as computational nodes, where certain variables are computed, namely fluid pressure or vapor partial pressure. Transfer phenomena are described by one-dimensional approximations at the discrete pore level. Based on the mass balance at each node, two linear systems are formed to be solved numerically, in order to obtain partial vapor pressure in each gas pore (and in the boundary layer) and fluid pressure in each pore.

Correct determination of macroscopic parameters becomes the main problem to be solved. Through continuous advances in the imaging technology [34], as well as the use of methods of pore networks construction based on digital images of microstructures [35], it will only be a matter of time before these parameters are precisely determined based on the high performance pore network computations.

2. Materials and Methods

In this paper, a pore network model is used to study heat and mass transfer through a capillary-porous building material. In order to study temperature and moisture conditions of the capillary-porous material, a corresponding computational grid is formed, which is a system of rectangular channels, arranged in parallel to coordinate axes, and intersecting with each other.

Equivalent diameters of these channels correspond to the average pore diameter of the analyzed porous medium; a ratio of the total pore volume to the porous material volume corresponds to this material porosity.

2.1. Dispatch Model and Data

2.1.1. Computational Grid

One of the options to construct such a network is shown in Figure 3a–d.

The design model is based on a cubic element with s side. The pores are represented as intersecting square section channels. The side of the d_k square corresponds to the known equivalent pore diameter of the material. The side length s of a cubic element is calculated from the condition

$$s^3 \varepsilon = 3 s d_k^2 - 2 d_k^3 \tag{1}$$

where ε is the known material porosity, expressing a ratio of pore volume to the total volume of the porous material.

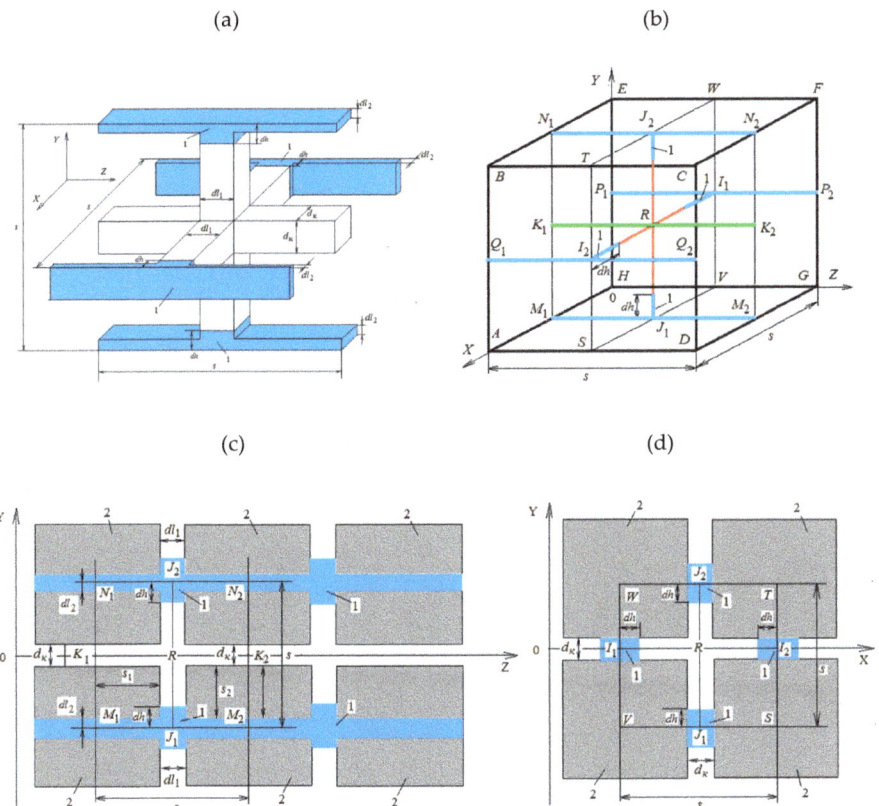

Figure 3. Cubic element of the computational scheme of the capillary-porous material network model: (**a**) volumetric image of a single computational cubic cell: (**b**) diagram of the lines of movement of material flows: (**c**) intersection of a cubic element by Y0Z plane: (**d**) intersection of a cubic element by Y0X plane.

The analyzed network model of heat and moisture transfer through a porous medium assumes that the most intensive vapor-air mixture (gas phase) and heat transfer occurs in the direction of $0Z$ axis through K_1K_2 channels (Figure 3b,c), arranged parallel to this axis. These channels in sections are squares with d_k side. Regarding heat and mass transfer through the building wall constructions, these channels are considered perpendicular to outer and inner surfaces of the building enclosure and connected to the inner and outer air media. According to this model, a liquid phase is arranged in the form of separate inclusions (mark 1 in Figure 3) in the network channels with I_1I_2; J_1J_2 axes are parallel to $0X$ and $0Y$ axes and perpendicular to K_1K_2 channels through which the gas phase is transferred. These inclusions of the liquid phase are in the form of rectangular columns. It is assumed that cross-sections of the channels, where liquid columns are arranged, are rectangles with d_k and dl_1 sides. It is assumed that columns with the liquid phase are interconnected by channels with M_1M_2, N_1N_2, P_1P_2 and Q_1Q_2 axes; they are parallel to the $0Z$ axis. The channels, connecting liquid columns, also contain a liquid phase. According to the assumed model, liquid evaporates or condenses on the column surfaces, occupying sections I_1I_2 and J_1J_2 channels (Figure 3).

As a result of evaporation or condensation, the liquid mass in these columns, as well as their height, can vary with time. The liquid mass in the column-connecting channels is considered constant with time. Width of specified channels with M_1M_2, N_1N_2, P_1P_2 and Q_1Q_2 axes corresponds to d_k value, and their height dl_2 is calculated from the minimum possible moisture content of the liquid phase $w_{l,min}$ in the material, corresponding to conditions of the analyzed problem, $w_{l,min}$ value is determined by the minimum relative air humidity φ_{min} in a porous material or in the external medium during the entire process of heat and mass transfer. This value is taken from the problem's initial or boundary conditions. In order to determine $w_{l,min}$ from φ_{min}, the sorption-desorption curve for an analyzed material should be used.

The liquid phase moisture content is considered as a ratio of the liquid mass in a certain volume of the porous material to this volume value. In the scope of considered cubic element, the moisture content is described by $w_{l\,min} = m_l/s^3$ expression, where ml is the liquid moisture mass, contained in this element. The liquid mass, contained in the considered channels with M_1M_2, N_1N_2, P_1P_2 and Q_1Q_2 axes, can be calculated as

$$m_{l,min} = 4sd_K dl_2 \rho_l.$$

This value can also be obtained from the following expression

$$m_{l,min} = w_{l,min}(\phi_{min})s^3.$$

By making the last two expressions equal, we can get the width of channels dl_2

$$dl_2 = \frac{w_{l,min}(\phi_{min})s^2}{4d_K \rho_l}.$$

With this configuration of the computational domain, the total pore volume in the considered cubic element is

$$V_p = d_K^2 s + 4dl_1 d_K \times \left(\frac{s}{2} - \frac{d_K}{2}\right) + 4(s - dl_1)d_K dl_2.$$

This value shall correspond to the specified material porosity ε. It follows from the condition (1) that

$$V_p = s^3 \varepsilon = 3sd_k^2 - 2d_k^3.$$

By making the last two expressions for V_p equal, we can get the width of channels dl_1, where liquid columns are arranged

$$dl_1 = \frac{d_K \times \left(\frac{s}{2} - \frac{d_K}{2}\right) - dl_2 \times s}{\left(\frac{s}{2} - \frac{d_K}{2} - dl_2\right)}.$$

This network model assumes that the heat and mass transfer processes proceed symmetrically relative to $ABCD$, $HEFG$, $BEFC$ and $AHGD$ planes. That is, there is no mass and heat transfer through these planes.

Intersections of these symmetry planes with the section, shown in Figure 3c, are represented by N_1N_2 and M_1M_2 segments, whereas intersections of symmetry planes with the section, shown in Figure 3d, correspond to WT, TS, SV and VW segments. The pore volume in a cubic element occupied by the liquid phase is

$$V_l = 4d_k \times dl_2 \times s + 4d_k \times dl_1 \times (dh - dl_2). \tag{2}$$

If the moisture content of the liquid phase in a porous material is equal to w_l, then its mass in the considered cubic element is

$$\delta m_l = V_l \times \rho_l = w_{l,0} \times s^3.$$

This equality, taking expression (2) into account, makes it possible to establish a relationship between the height of liquid columns dh and the material moisture content w_l

$$dh = \frac{w_l s^3 + 4\rho_l d_k dl_2 \times (dl_1 - s)}{4\rho_l d_k dl_1}. \tag{3}$$

2.1.2. Transfer Model

In order to study the behavior of moisture content and temperature of the porous material with time, the mass and heat balance equations are formulated for the considered cubic elements, arranged sequentially in the direction of a $0Z$ axis. The balance equations are formulated for discrete instants of time τ_k with $\Delta\tau$ interval.

The mass transfer in a gas phase, i.e., in a mixture of dry air and water vapor, occurs mainly in a channel with $K_1 K_2$ axes by molecular diffusion and filtration. Mass transfer by the diffusion occurs due to mass concentration gradients (partial density) of dry air and vapor in a gas mixture, and it is described by Fick's law:

$$j_{a,dif} = -D_{va}\frac{\partial \rho_a}{\partial z} \tag{4}$$

$$j_{v,dif} = -D_{va}\frac{\partial \rho_a}{\partial z} \tag{5}$$

where ρ_v, [kg/m^3] is partial density of water vapor in a mixture; ρ_a, [kg/m^3] is partial density of dry air in a mixture; $j_{v,dif}$, [kg/(m^{2}s)] is vapor flow density due to diffusion; $j_{a,dif}$, [kg/(m^{2}s)] is dry air flow density due to diffusion; D_{va}, [m^2/s] is the diffusion coefficient of water vapor and dry air in a gas mixture.

Density values of dry air and water vapor are calculated according to the ideal gas state equations;

$$\rho_a = \frac{p_a}{R_a T_g} \tag{6}$$

$$\rho_v = \frac{p_v}{R_a T_g} \tag{7}$$

where p_a, p_v, [Pa] is partial pressure of dry air and water vapor in a mixture; R_a, R_v, [J/(kg·K)] are gas constants of dry air and water vapor; T_g, [K] is gas mixture temperature. Besides, the gas medium (vapor-air mixture) transfer also occurs due to filtration.

Density values of vapor and air flows due to filtration are described by the Darcy equations;

$$j_{a,fil} = -\rho_a \frac{K_g}{\mu_g}\frac{\partial p_g}{\partial z} \tag{8}$$

$$j_{v,fil} = -\rho_v \frac{K_g}{\mu_g}\frac{\partial p_g}{\partial z} \tag{9}$$

where K_g, [m^2] is permeability coefficient of the porous material for a gas medium; μ_g, [Pa·s] is dynamic viscosity coefficient of a gas medium; $p_g = p_a + p_v$, [Pa] is vapor-air medium pressure.

The gas phase in a cubic element occupies the space of a channel with K_1K_2 axis, as well as part of the channel volumes with I_1I_2 and J_1J_2 axes (Figure 3). The volume, occupied by a gas phase, is calculated from the following expression:

$$V_{gi}^k = d_K^2 s + 4\left(\frac{s}{2} - \frac{d_K}{2} - dh_i^k\right) d_K dl_1 \qquad (10)$$

where dh_i^k is liquid column height in the i-th element at τ_k instant of time.

The balance equation of dry air mass in a cubic element with i number for τ_k instant of time is derived from the condition that air enters the considered element with diffusion $J^+_{a,dif}$ and filtration $J^+_{a,fil}$ flows from an adjacent element with $i-1$ number through a surface with d_K^2 area, and it is transferred to the next adjacent element with $i+1$ number flows $J^-_{a,dif}$ and $J^-_{a,fil}$. In order to derive this equation, in expressions (4), (8) describing air flows by diffusion and filtration, the derivatives with respect to z variable are replaced by finite differences.

Dry Air Transport Dodel

Taking expression (6) into account, this equation is represented as:

$$\frac{p_{a,i}^k}{R_a T_{g,i}^k} V_{gi}^k - \frac{p_{a,i}^{k-1}}{R_a T_{g,i}^{k-1}} V_{gi}^{k-1} = \left(J^+_{a,dif} - J^-_{a,dif} + J^+_{a,fil} - J^-_{a,fil}\right) d_K^2 \Delta\tau \qquad (11)$$

$$J^-_{a,dif} = -\frac{D_{va,i+1/2}}{s}\left(\frac{p_{a,i+1}^k}{R_a T_{g,i+1}^k} - \frac{p_{a,i}^k}{R_a T_{g,i}^k}\right)$$

$$J^+_{a,fil} = -\frac{p_{a,i-1/2}^k}{s}\frac{K_g}{\mu_g}\left(p_{a,i}^k + p_{v,i}^k - p_{a,i-1}^k - p_{v,i-1}^k\right)$$

$$J^-_{a,fil} = -\frac{p_{a,i+1/2}^k}{s}\frac{K_g}{\mu_g}\left(p_{a,i+1}^k + p_{v,i+1}^k - p_{a,i}^k - p_{v,i}^k\right).$$

This is the conservation equation for the local dry air mass. The left side of the equation is the mass difference in an elementary cubic cell between two successive points in time (through a time step), obtained from the law for an ideal gas. The right-hand side is recorded for the same times and consists of the difference in mass flows due to diffusion due to the concentration gradient and mass flow due to filtration due to the total pressure gradient. These two effects on the right side of expression (11) are not opposite to each other, but complement each other. Many researchers use this approach.

In this discrete equation, the values with the i index describe gas medium parameters in the considered an element of the porous material. Formally, it is considered that they refer to R node, located in the center of this element (Figure 3). Values with fractional indices are calculated as arithmetic (or weighted) mean values related to adjacent elements. Values with k index refer to the current moment of time, and those with the $k-1$ index to the previous one.

Water Vapor Transfer Model

The mass balance equation for water vapor is also based on the condition that vapor transfer through a cubic element occurs by diffusion and filtration in the direction of $0Z$ axis. Diffusion and filtration water vapor flows are described by expressions (5) and (9), where the derivatives are replaced by finite differences.

Besides, it is considered that water vapor, evaporated from liquid column surfaces enters the gas medium with a diffusion flow $J^+_{l\_v,dif}$ through $I_1I_2; J_1J_2$ channels.

The vapor mass balance equation, considering expression (7), is written as:

$$\frac{p_{v,i}^k}{R_v T_{g,i}^k} V_{gi}^k - \frac{p_{v,i}^{k-1}}{R_v T_{g,i}^{k-1}} V_{gi}^{k-1} = \left(J^+_{v,dif} - J^-_{v,dif} + J^+_{v,fil} - J^-_{v,fil}\right) d_K^2 \Delta\tau + \qquad (12)$$
$$+4 J_{l\_v,dif} dl_1 d_K \Delta\tau,$$

where

$$J^+_{v,dif} = -\frac{D_{va,i-1/2}}{s}\left(\frac{p^k_{v,i}}{R_v T^k_{g,i}} - \frac{p^k_{v,i-1}}{R_v T^k_{g,i-1}}\right);$$

$$J^-_{v,dif} = -\frac{D_{va,i+1/2}}{s}\left(\frac{p^k_{v,i+1}}{R_v T^k_{g,i+1}} - \frac{p^k_{v,i}}{R_v T^k_{g,i}}\right);$$

$$J^+_{v,fil} = -\frac{\rho^k_{v,i-1/2}}{s}\frac{K_g}{\mu_g}\left(p^k_{a,i} + p^k_{v,i} - p^k_{a,i-1} - p^k_{v,i-1}\right);$$

$$J^-_{v,fil} = -\frac{\rho^k_{v,i+1/2}}{s}\frac{K_g}{\mu_g}\left(p^k_{a,i+1} + p^k_{v,i+1} - p^k_{a,i} - p^k_{v,i}\right);$$

$$J_{l\_v,dif} = -\frac{D_{va,i}}{\frac{s}{2} - dh^k_i}\left(\frac{p^k_{v,i}}{R_v T^k_{g,i}} - \frac{p^k_{v\_l,i}}{R_v T^k_{g\_l,i}}\right) + \frac{p^k_{v\_l,i}}{R_v T^k_{g\_l,i}}u_s;$$

$\left(\frac{s}{2} - dh^k_i\right)$ is a distance from the surface of liquid columns to R point; u_s is Stefan's speed; $p^k_{v\_l,i}$ is the partial pressure of water vapor directly above the surface of liquid columns; $T^k_{g\_l,i}$ is the temperature of the liquid column surface, where liquid is evaporated from.

Liquid (Water) Transfer Model

Liquid phase transfer in the channels with M_1M_2, N_1N_2, P_1P_2 and Q_1Q_2 axes occurs due to filtration, resulting from the action of pressure gradient in a liquid medium. This filtration flow is described by the Darcy equation

$$j_l = -\rho_l \frac{K_l}{\mu_l}\frac{\partial p_l}{\partial z} \tag{13}$$

where j_l [kg/(m²s)] is density of the filtration fluid flow; p_l [Pa] is pressure in a liquid phase; μ_l [Pa·s] is the dynamic coefficient of medium liquid viscosity; [Pa·s] is the dynamic coefficient of medium liquid viscosity; ρ_l [kg/m³] is liquid density; K_l [m²] is the permeability coefficient of the porous material for a liquid medium. Pressure in a liquid phase is defined as the difference between a vapor-gas medium and capillary pressure:

$$p_l = p_g - p_c.$$

Considering this expression, Equation (13) can be written as

$$j_l = \rho_l \frac{K_l}{\mu_l}\frac{\partial p_c}{\partial z} \tag{14}$$

since it can be assumed that $\frac{\partial p_g}{\partial z} \ll \frac{\partial p_c}{\partial z}$.

Capillary pressure p_c depends on the specific moisture content w_l. In this regard, derivative $\frac{\partial p_c}{\partial z}$ in the expression (16) is replaced by $\frac{\partial p_c}{\partial z} = \frac{dp_c}{dw_l}\frac{dw_l}{dh}\frac{\partial h}{\partial z}$. Derivative $\frac{dp_c}{dw_l}$ is determined from the experimental dependence of capillary pressure p_c on the specific moisture content w_l, derivative $\frac{dw_l}{dh}$ is calculated from the expression (3):

$$\frac{dw_l}{dh} = \frac{4dl_1 d_K \rho_l}{s^3} = C_h$$

Accordingly, the mass balance equation for a liquid phase is derived

$$\rho_l\left(dh^k_i - dh^{k-1}_i\right)dl_1 d_K = \left(J^+_{l,fil} - J^-_{l,fil}\right)dl_2 d_K \Delta\tau - J_{l\_v,dif}dl_1 d_K \Delta\tau; \tag{15}$$

where

$$J^+{}_{l,fil} = \frac{\rho_l}{\mu_l} C_h \left(K_l \frac{dp_c}{dw_l} \right)\bigg|_{i-1/2} \frac{h_i^k - h_{i-1}^k}{s};$$

$$J^-{}_{l,fil} = \frac{\rho_l}{\mu_l} C_h \left(K_l \frac{dp_c}{dw_l} \right)\bigg|_{i+1/2} \frac{h_{i+1}^k - h_i^k}{s}.$$

Equation (17), as well as Equation (13) for water vapor consider the diffusion transfer $J_{l\_v,dif}$ of evaporated moisture from liquid column surfaces into a gas phase.

The partial pressure of water vapor above the surface of liquid columns is calculated as

$$p_{v\_l}{}_{,i}^k = p_{sut}\left(T_{g\_l}{}_{,i}^k\right) \cdot \varphi\left(w_{l,i}^k, T_{g\_l}{}_{,i}^k\right),$$

where $p_{sut}\left(T_{g\_l}{}_{,i}^k\right)$ is saturation pressure, corresponding to the surface temperature of liquid columns; $\varphi\left(w_{l,i}^k, T_{g\_l}{}_{,i}^k\right)$ is relative air humidity, corresponding to specific moisture content $w_{l,i}^k$. This dependence is determined from the sorption-desorption isotherm for a specified material.

Model of Heat Transfer in a Vapor-Air Medium

The energy conservation equation for volume $V_{g,i}^k$ of the vapor-air mixture is based on the condition that heat enters this volume by convection Q_{g_conv} and heat conductivity Q_{g_cond}. The heat convective flows Q_{g_conv} are created by diffusion and filtration flows of dry air and water vapor.

In addition, the heat $Q_{l\_g\_conv}$ is transferred by convection into a gas medium with moisture flow, evaporated from liquid column surfaces. The heat flow with heat conductivity that Q_{g_cond} generates is due to the presence of a temperature gradient in a gas medium along the $0Z$ axis. By means of heat conductivity, the heat $Q_{l\_g\_cond}$ also enters the considered volume from the surface of liquid columns, resulting from temperature differences between a gas medium and a liquid phase. Besides, the heat $Q_{s\_g\_cond} = Q_{s1\_g\_cond} + Q_{s2\_g\_cond}$ centers a gas medium from the surfaces of pore walls by means of heat conductivity. A block diagram of the movement of heat and material flows (and their corresponding designations) for the central nodal part of a single elementary cubic element of material, which is shown in Figure 3c, which in an enlarged form is shown in Figure 4.

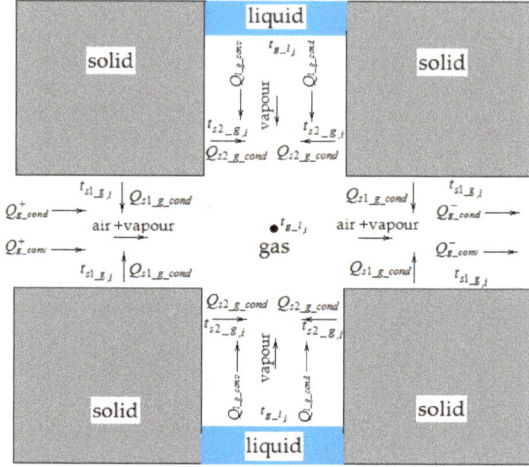

Figure 4. Scheme of heat transfer for the gas phase in an elementary cubic element.

Considering the above, the heat balance equation for a gas medium is derived:

$$V_{gi}^k \left(I_{ai}^k + I_{vi}^k\right) = V_{gi}^{k-1}\left(I_{ai}^{k-1} + I_{vi}^{k-1}\right) + Q^+{}_{g\_conv} - Q^-{}_{g\_conv} + Q^+{}_{g\_cond} - Q^-{}_{g\_cond} + Q_{l\_g\_conv} + Q_{l\_g\_cond} + Q_{s1\_g\_cond} + Q_{s2\_g\_cond};$$

where

$$Q^+{}_{g\_conv} = \left(J^+{}_{a,dif} + J^+{}_{a,fil}\right) Ia^k_{i-1/2} d_K^2 \Delta\tau + \left(J^+{}_{v,dif} + J^+{}_{v,fil}\right) Iv^k_{i-1/2} d_K^2 \Delta\tau;$$

$$Q^-{}_{g\_conv} = \left(J^-{}_{a,dif} + J^-{}_{a,fil}\right) Ia^k_{i+1/2} d_K^2 \Delta\tau + \left(J^-{}_{v,dif} + J^-{}_{v,fil}\right) Iv^k_{i+1/2} d_K^2 \Delta\tau;$$

$$Q^+{}_{g\_cond} = q^+{}_g d_K^2 \Delta\tau;\ Q^-{}_{g\_cond} = q^-{}_g d_K^2 \Delta\tau;$$

$$Q_{l\_g\_conv} = 4 J_{l\_v,dif} I_{l\_v i}^k dl_1 d_K \Delta\tau;\ Q_{l\_g\_cond} = 4 q^+{}_{l\_g} dl_1 d_K \Delta\tau;$$

$$Q_{s1\_g\_cond} = 4 q^+{}_{s1\_g} (s - dl_1) d_K \Delta\tau;\ Q_{s2\_g\_cond} = 4 q^+{}_{s2\_g} f_{s2\_g} \Delta\tau + 4 q^+{}_{s3\_g} f_{s3\_g} \Delta\tau.$$

After substituting these expressions into the heat balance equation for gas, we get:

$$\begin{aligned}
V_{gi}^k \left(I_{ai}^k + I_{vi}^k\right) &= V_{gi}^{k-1}\left(I_{ai}^{k-1} + I_{vi}^{k-1}\right) + \\
&+ \left(J^+{}_{a,dif} + J^+{}_{a,fil}\right) Ia^k_{i-1/2} d_K^2 \Delta\tau - \left(J^-{}_{a,dif} + J^-{}_{a,fil}\right) Ia^k_{i+1/2} d_K^2 \Delta\tau + \\
&+ \left(J^+{}_{v,dif} + J^+{}_{v,fil}\right) Iv^k_{i-1/2} d_K^2 \Delta\tau - \left(J^-{}_{v,dif} + J^-{}_{v,fil}\right) Iv^k_{i+1/2} d_K^2 \Delta\tau + \\
&+ 4 J_{l\_v,dif} I_{l\_v i}^k dl_1 d_K \Delta\tau + q^+{}_g d_K^2 \Delta\tau - q^-{}_g d_K^2 \Delta\tau \\
&+ 4 q^+{}_{l\_g} dl_1 d_K \Delta\tau + 4 q^+{}_{s1\_g} (s - dl_1) d_K \Delta\tau + 4 q^+{}_{s2\_g} f_{s2\_g} \Delta\tau + 4 q^+{}_{s3\_g} f_{s3\_g} \Delta\tau
\end{aligned} \qquad (16)$$

where

$$I_{ai}^k = C_a t_{gi}^k \frac{p a_i^k}{R_a T_{gi}^k};\ I_{vi}^k = \left[C_w t_n\left(p_{vi}^k\right) + r_v + C_v\left(t_{gi}^k - t_n\left(p_{vi}^k\right)\right)\right] \frac{p_{vi}^k}{R_v T_{gi}^k};$$

$$I_{l\_v i}^k = C_w t_{g\_l,i}^k + r_v;\ q^+{}_g = -\lambda_g \frac{t_{gi}^k - t_{gi-1}^k}{s};\ q^-{}_g = -\lambda_g \frac{t_{gi+1}^k - t_{gi}^k}{dz};$$

$$q^+{}_{l\_g} = -\lambda_g \frac{t_{g,i}^k - t_{g\_l,i}^k}{\left(\frac{s}{2} - dh_i^k\right)};\ q^+{}_{s1\_g} = -\lambda_g \frac{t_{g,i}^k - t_{s1\_g,i}^k}{d_K/2};\ q^+{}_{s2\_g} = -\lambda_g \frac{t_{g,i}^k - t_{s2\_g,i}^k}{dl_1/2};$$

$$q^+{}_{s3\_g} = -\lambda_g \frac{t_{g,i}^k - t_{s2\_g,i}^k}{d_K/2};\ f_{s2\_g} = 2 d_K \left(\frac{s}{2} - \frac{d_K}{2} - dh_i^k\right);\ f_{s3\_g} = 2 dl_1 \left(\frac{s}{2} - \frac{d_K}{2} - dh_i^k\right).$$

$t_{s1\_g,i}^k$ is temperature [°C] of the channel wall surfaces with $K_1 K_2$ axis, which is in contact with a vapor-gas medium; $t_{s2\_g,i}^k$ is the temperature of channel wall surfaces with $I_1 I_2$ and $J_1 J_2$ axes, which are in contact with a vapor-gas medium; $t_n\left(p_{vi}^k\right)$ is saturation temperature, corresponding to vapor pressure; p_{vi}^k; C_a; C_v; C_w, [J/(kg·K)] are specific heat capacity values of dry air, water vapor and water; r_v, [J/kg] is specific heat of vapor formation; λ_g, [W/(m·K)] is the heat conductivity coefficient of a vapor-gas mixture; f_{s2_g}; f_{s3_g} are contact surfaces of a vapor-gas mixture with channel walls with $I_1 I_2$; $J_1 J_2$ axes $I_1 I_2$; $J_1 J_2$.

Model of Heat Transfer in the Liquid Phase

The energy conservation equations for a liquid phase are derived for liquid volume $V_l = d_K dl_2 s + \left(dh_i^k - dl_2\right) d_K dl_1$, including channel volumes, dl_2, high, containing a constant liquid volume, and volumes of liquid columns, $dh_i^k - dl_2$ varying with time. If the liquid temperature value in J_1; J_2; I_1; I_2 nodes (Figure 1) in the design element with i number is $t_{w,i}^k$, then the heat content in this liquid volume at the time step k is calculated from the expression $Ql_i^k = C_w \rho_w t_{w,i}^k \left[d_K dl_2 s + \left(dh_i^k - dl_2\right) d_K dl_1\right]$.

Through channels with $M_1 M_2$, $N_1 N_2$, $P_1 P_2$ $Q_1 Q_2$ axes, where the liquid fraction is located, the heat transfer is performed by means of heat conductivity due to the temperature gradient, as well as by convection with filtration liquid flows.

From the surfaces of liquid columns, heat is removed from the considered volume by means of heat conductivity and convection with liquid flow, evaporating from the column surface and transferred into a vapor-air mixture. Heat is also transferred from liquid to channel walls by M_1M_2, N_1N_2, P_1P_2 and Q_1Q_2 axes, as well as to channel walls with J_1J_2 and I_1I_2 axes, where the liquid columns are located. Thus, the energy conservation equation for the liquid fraction is represented as:

$$C_w \rho_w tw_{l,i}^k V_{l,i}^k = C_w \rho_w tw_{l,i}^{k-1} V_{l,i}^{k-1} - q^-{}_{l\_g} dl_1 d_K \Delta\tau - J_{l\_v,dif} C_w tg_{\_l,i}^k dl_1 d_K \Delta\tau \\ + (q^+{}_w - q^-{}_w) d_K dl_2 \Delta\tau + \left(J^+{}_{l,fil} C_w tw_{i-1/2}^k - J^-{}_{l,fil} C_w tw_{i+1/2}^k \right) dl_2 d_K \Delta\tau - \\ - q^-{}_{l\_s1} f_{s1\_l} \Delta\tau + q^-{}_{l\_s2} f_{s2\_l} \Delta\tau + q^-{}_{l\_s3} f_{s3\_l} \Delta\tau \quad (17)$$

where

$q^-{}_{l\_g} = -\lambda_w \frac{tg_{\_l,i}^k - tw_{i}^k}{dh_i^k}$; $q^+{}_w = -\lambda_w \frac{tw_i^k - tw_{i-1}^k}{s}$; $q^-{}_w = -\lambda_w \frac{tw_{i+1}^k - tw_i^k}{s}$;

$q^-{}_{l\_s1} = -\lambda_w \frac{tw-s1_{\_i}^k - tw_{\_i}^k}{dl_2}$; $q^-{}_{l\_s2} = -\lambda_w \frac{tw_i^k - tw-s2_{\_i}^k}{dl_1/2}$; $q^-{}_{l\_s3} = -\lambda_w \frac{tw_i^k - tw-s2_{\_i}^k}{d_K/2}$;

$f_{s1\_l} = (2dl_2 + d_K)(s - dl_1)$; $f_{s2\_l} = 2d_K\left(dh_i^k - dl_2\right)$; $f_{s3\_l} = 2dl_1\left(dh_i^k - dl_2\right)$.

$tw_{-s1,i}^k$ is the temperature ([°C]) of channel wall surfaces with M_1M_2; N_1N_2; P_1P_2; Q_1Q_2 axes, which are in contact with a liquid phase; $tw_{-s2,i}^k$ is temperature of channel wall surfaces with I_1I_2; J_1J_2 axes, which are in contact with a liquid medium; λ_w, W/(m K) is liquid heat conductivity coefficient; f_{s1_l} is a contact surface of a liquid phase with channel walls, through which the liquid is filtered f_{s2_l}; f_{s3_l}- are contact surfaces of a vapor-gas mixture with channel walls, where the liquid column is located.

Heat Transfer Model in a Solid Structure

The energy conservation equation for a solid fraction of the considered element, occupying volume $V_s = s^3(1 - \varepsilon)$, is derived taking into account the fact that heat transfer occurs along a solid body in the direction of 0Z axis by means of heat conductivity.

The heat flow by means of heat conductivity enters the cubic element and leaves it through the face with the area of $f_s = s^2 - d_K^2 - 4d_K dl_2$. On the surfaces of channel walls with the K_1K_2 axis, the heat exchange of a solid fraction with a gas medium occurs. On channel walls with I_1I_2 and J_1J_2 axes, the heat exchange occurs with a gas phase, present in these channels, as well as with liquid columns. On channel walls with M_1M_2, N_1N_2, P_1P_2 and Q_1Q_2 axes, there is a heat exchange of solid phase with a liquid medium.

Thus, the energy conservation equation for the solid fraction is represented as:

$$C_s \rho_s ts_{,i}^k V_s = C_s \rho_s ts_{,i}^{k-1} V_s + q^+{}_s f_s \Delta\tau - q^-{}_s f_s \Delta\tau - 4q^-{}_{s1\_g}(s-dl_1)d_K \Delta\tau + \\ + 4q^+{}_{s1\_l} f_{s1\_l} \Delta\tau - 4q^+{}_{l\_s2} f_{s2\_l} \Delta\tau - 4q^+{}_{l\_s3} f_{s3\_l} \Delta\tau - \\ - 4q^-{}_{s2\_g} f_{s2\_g} \Delta\tau - 4q^-{}_{s3\_g} f_{s3\_g} \Delta\tau \quad (18)$$

where

$q^+{}_s = -\lambda_s \frac{ts_i^k - ts_{i-1}^k}{s}$; $q^-{}_s = -\lambda_s \frac{ts_{i+1}^k - ts_i^k}{s}$; $q^-{}_{s1\_g} = -\lambda_s \frac{ts1_{\_g,i}^k - ts_{\_i}^k}{s_2/2}$; $q^+{}_{s1\_l} = -\lambda_s \frac{ts_{\_i}^k - tw-s1_{\_i}^k}{s_2/2}$;

$s_2 = 0,5(s - d_K) - dl_2$;

$q^+{}_{l\_s2} = -\lambda_s \frac{tw-s2_{\_i}^k - ts_{\_i}^k}{s_{1,1}/2}$; $q^+{}_{l\_s3} = -\lambda_s \frac{tw-s2_{\_i}^k - ts_{\_i}^k}{s_{1,2}/2}$;

$q^-{}_{s2\_g} = -\lambda_s \frac{ts2_{\_g,i}^k - ts_{\_i}^k}{s_{1,1}/2}$; $q^-{}_{s3\_g} = -\lambda_s \frac{ts2_{\_g,i}^k - ts_{\_i}^k}{s_{1,2}/2}$;

$s_{1,1} = 0,5(s - dl_1)$; $s_{1,2} = 0,5(s - d_K)$.

The temperature-moisture state of a capillary-porous material is described by a system of equations for the mass and energy conservation: (11); (12); (15); (16); (17); (18). This system of equations is written for all cubic elements with the numbers $I = 1...N$.

Its solution makes it possible to calculate the values of network functions describing: partial pressure of dry air $p_{a,i}^k$; partial pressure of water vapor $p_{v,i}^k$; height of liquid columns dh_i^k; vapor-air mixture temperature tg_i^k; liquid phase temperature $tw_{,i}^k$ and solid fraction temperature of a porous material $t_{s,i}^k$. Except for indicated values, this system of equations also contains: liquid column surface temperature $t_{g\_l,i}^k$; surface temperature of channel walls with K_1K_2 axis, which is in contact with a vapor-gas medium $t_{s1\_g,i}^k$; surface temperature of channel walls with I_1I_2; J_1J_2 axes, which are in contact with a vapor-gas medium $t_{s2\_g,i}^k$; surface temperature of channel walls with M_1M_2; N_1N_2; P_1P_2; Q_1Q_2 axes, which are in contact with a liquid phase $t_{w-s1,i}^k$ and surface temperature of channel walls with I_1I_2 and J_1J_2 axes, which are in contact with a liquid medium $t_{w-s2,i}^k$. In order to determine specified temperature values on the medium contact surfaces, the matching conditions are used.

2.1.3. The Matching Conditions on the Surfaces

The matching conditions on the surfaces of liquid columns, where a vapor-air mixture contacts with the liquid, and from which evaporation (condensation) occurs, are as follows:

$$q^-{}_{l\_g} = r_v J_{l\_v,dif} + q^+{}_{l\_g}$$

or

$$-\lambda_w \frac{t_{g\_l,i}^k - t_{w,i}^k}{dh_i^k} = r_v J_{l\_v,dif} - \lambda_g \frac{t_{g,i}^k - t_{g\_l,i}^k}{0,5s - dh_i^k}.$$

The value is determined from this expression: $t_{g\_l,i}^k$:

$$t_{g\_l,i}^k = \frac{\frac{\lambda_w}{dh_i^k}}{\left(\frac{\lambda_g}{0,5s-dh_i^k} + \frac{\lambda_w}{dh_i^k}\right)} t_{w,i}^k + \frac{\frac{\lambda_g}{0,5s-dh_i^k}}{\left(\frac{\lambda_g}{0,5s-dh_i^k} + \frac{\lambda_w}{dh_i^k}\right)} t_{g,i}^k - \frac{r_v J_{l\_v,dif}}{\left(\frac{\lambda_g}{0,5s-dh_i^k} + \frac{\lambda_w}{dh_i^k}\right)}.$$

The matching condition on channel walls with M_1M_2; N_1N_2; P_1P_2; Q_1Q_2 axes, which are in contact with a liquid phase, are represented as:

$$q^-{}_{l\_s1} = q^+{}_{s1\_l}$$

or

$$-\lambda_w \frac{t_{w-s1,i}^k - t_{w,i}^k}{dl_2} = -\lambda_s \frac{t_{s,i}^k - t_{w-s1,i}^k}{0,5s_2}.$$

This expression determines $t_{w-s1,i}^k$

$$t_{w-s1,i}^k = \frac{\frac{\lambda_w}{dl_2}}{\left(\frac{\lambda_s}{0,5s_2} + \frac{\lambda_w}{dl_2}\right)} t_{w,i}^k + \frac{\frac{\lambda_s}{0,5s_2}}{\left(\frac{\lambda_s}{0,5s_2} + \frac{\lambda_w}{dl_2}\right)} t_{s,i}^k.$$

The matching condition on channel walls with K_1K_2 axis, where the heat exchange of a vapor-gas mixture with a solid phase of the porous material occurs, is as follows:

$$q^+{}_{s1\_g} = q^-{}_{s1\_g}$$

or

$$-\lambda_g \frac{t_{g,i}^k - t_{s1\_g,i}^k}{0,5d_K} = -\lambda_s \frac{t_{s1\_g,i}^k - t_{s,i}^k}{0,5s_2}.$$

This equation determines the contact surface temperature of channel walls with K_1K_2 axis with a vapor-air medium

$$t_{s1\_g,i}^k = \frac{\frac{\lambda_s}{s_2}}{\left(\frac{\lambda_g}{d_K} + \frac{\lambda_s}{s_2}\right)} t_{s,i}^k + \frac{\frac{\lambda_g}{d_K}}{\left(\frac{\lambda_g}{d_K} + \frac{\lambda_s}{s_2}\right)} t_{g,i}^k.$$

In order to determine the surface temperature $t_{s2\_g,i}^k$ of channel walls with I_1I_2; J_1J_2 axes, which are in contact with a vapor-gas medium, the matching conditions are represented as

$$q^+{}_{s2\_g}f_{s2\_g} + q^+{}_{s3\_g}f_{s3\_g} = q^-{}_{s2\_g}f_{s2\_g} + q^-{}_{s3\_g}f_{s3\_g}$$

or with consideration of the above expressions

$$-\lambda_g \frac{t_{g,i}^k - t_{s2\_g,i}^k}{dl_1/2} 2d_K\left(\frac{s}{2} - \frac{d_K}{2} - dh_i^k\right) - \lambda_g \frac{t_{g,i}^k - t_{s2\_g,i}^k}{d_K/2} 2dl_1\left(\frac{s}{2} - \frac{d_K}{2} - dh_i^k\right) =$$
$$= -\lambda_s \frac{t_{s2\_g,i}^k - t_{s,i}^k}{s_{1,1}/2} 2d_K\left(\frac{s}{2} - \frac{d_K}{2} - dh_i^k\right) - \lambda_s \frac{t_{s2\_g,i}^k - t_{s,i}^k}{s_{1,2}/2} 2dl_1\left(\frac{s}{2} - \frac{d_K}{2} - dh_i^k\right).$$

From the presented expression, it follows that

$$t_{s2\_g,i}^k = \frac{\lambda_s\left(\frac{d_K}{s_{1,1}} + \frac{dl_1}{s_{1,2}}\right)}{\left(\lambda_g\left(\frac{d_K}{dl_1} + \frac{dl_1}{d_K}\right) + \lambda_s\left(\frac{d_K}{s_{1,1}} + \frac{dl_1}{s_{1,2}}\right)\right)} t_{s,i}^k + \frac{\lambda_g\left(\frac{d_K}{dl_1} + \frac{dl_1}{d_K}\right)}{\left(\lambda_g\left(\frac{d_K}{dl_1} + \frac{dl_1}{d_K}\right) + \lambda_s\left(\frac{d_K}{s_{1,1}} + \frac{dl_1}{s_{1,2}}\right)\right)} t_{g,i}^k.$$

The surface temperature $t_{w-s2,i}^k$ of channel walls with I_1I_2; J_1J_2 axes, which are in contact with a liquid medium, is determined using the matching condition

$$q^+{}_{l\_s2}f_{s2\_l} + q^+{}_{l\_s3}f_{s3\_l} = q^-{}_{l\_s2}f_{s2\_l} + q^-{}_{l\_s3}f_{s3\_l}.$$

which, taking above expressions into account, is represented as

$$-\lambda_s \frac{t_{w-s2,i}^k - t_{s,i}^k}{s_{1,1}/2} 2d_K\left(dh_i^k - dl_2\right) - \lambda_s \frac{t_{w-s2,i}^k - t_{s,i}^k}{s_{1,2}/2} 2dl_1\left(dh_i^k - dl_2\right) =$$
$$= -\lambda_w \frac{t_{w,i}^k - t_{w-s2,i}^k}{dl_1/2} 2d_K\left(dh_i^k - dl_2\right) - \lambda_w \frac{t_{w,i}^k - t_{w-s2,i}^k}{d_K/2} 2dl_1\left(dh_i^k - dl_2\right).$$

From this expression, it follows that

$$t_{w-s2,i}^k = \frac{\lambda_s\left(\frac{d_K}{s_{1,1}} + \frac{dl_1}{s_{1,2}}\right)}{\left(\lambda_w\left(\frac{d_K}{dl_1} + \frac{dl_1}{d_K}\right) + \lambda_s\left(\frac{d_K}{s_{1,1}} + \frac{dl_1}{s_{1,2}}\right)\right)} t_{s,i}^k + \frac{\lambda_w\left(\frac{d_K}{dl_1} + \frac{dl_1}{d_K}\right)}{\left(\lambda_w\left(\frac{d_K}{dl_1} + \frac{dl_1}{d_K}\right) + \lambda_s\left(\frac{d_K}{s_{1,1}} + \frac{dl_1}{s_{1,2}}\right)\right)} t_{w,i}^k.$$

2.2. Condition for Solving Equations

In order to solve the problem of heat and mass transfer dynamics in the considered formulation, the initial and boundary conditions for presented equations should be formulated. The initial distribution of temperature t_0 and moisture content $w_{l,0}$ over the material thickness can be set as initial conditions. The boundary conditions shall reflect ambient temperature t_∞ and a certain indicator of its moisture condition: relative air humidity φ_∞ or partial pressure of water vapor $p_{v,\infty}$ or its concentration $\rho_{v,\infty}$ in air. Also, the total pressure of a vapor-air mixture outside the material $p_{g,\infty} = p_{a,\infty} + p_{v,\infty}$, should be set; it usually corresponds to the atmospheric pressure.

3. Results

As an example, the change in temperature and moisture condition in time of a porous material, $Z = 0.1$ [m] thick was analyzed. Its porosity is $\varepsilon = 0.157$. Thermophysical properties of the considered material correspond to properties of a ceramic brick. Permeability coefficient for gaseous medium $K_g = 2.2 \times 10^{-13}$, [m^2]. For the dependences of the permeability coefficients and capillary pressure for the liquid in the material on the moisture content, the data given in [9] were used. Note that the values of capillary pressure and the coefficient of permeability of a liquid in a material depend significantly on its moisture content. For the considered range of changes in moisture content $w = 3...60$ [kg/m^3] capillary pressure, respectively, varied within $p_c = 9.5 \times 10^6...0.1 \times 10^6$ [Pa], and the ratio of the permeability coefficient (for liquid) to its dynamic coefficient viscosity varied in the range $K_l/\mu_l = 4.0 \times 10^{-16}...6.7 \times 10^{-11}$ [m^2] [9]. The heat capacity and thermal conductivity coefficients for each phase were chosen to be constant and, accordingly, equal: $C_a = 1006.43$; $C_v = 1875.2$; $C_w = 4183$, [J/(kg·K)] and $\lambda_g = 0.0259$; $\lambda_w = 0.612$; $\lambda_s = 0.7$, [W/(m·K)]. The specific heat of the liquid-vapor phase transition is $r_v = 2.260 \cdot 10^6$, [J/kg], and the diffusion coefficient of vapor in air is $D_{va} = 2.31 \times 10^{-5}$, [m^2/s].

3.1. Evaporation Processes

At the initial time, the material moisture content is $w_{l,0} = 60$ [kg/m^3]. This value produces half of the maximum possible moisture content in the material, at which time all pores are filled with liquid. At the specified moisture content, the relative air humidity inside the material is practically equal to one. The initial material temperature is 20 [°C]. The material is placed in an air medium, its temperature is also 20 [°C], the relative humidity is $\varphi = 0.6$. At this point, $\rho_{v,\infty} = 0.0104$ [kg/m^3].

The calculation results of the variation with time in temperature and moisture conditions of the capillary-porous material for these conditions are shown in Figure 5.

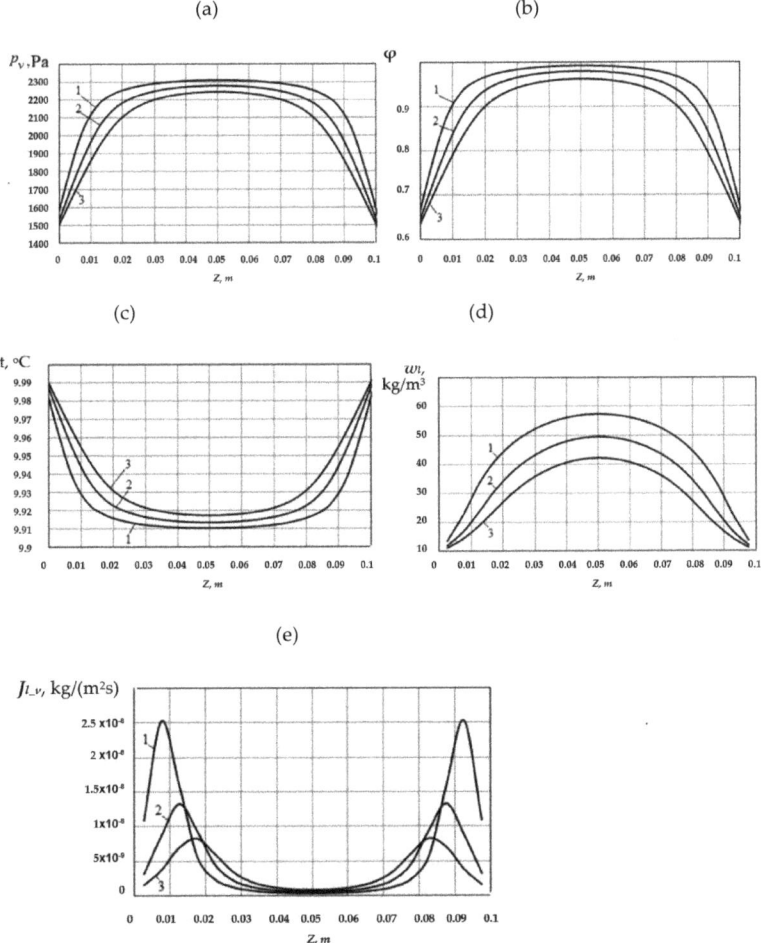

Figure 5. Distribution of the vapor partial pressure over the material thickness (**a**), relative air humidity (**b**), temperature (**c**), moisture content (**d**), and vapor flow density from the liquid column surfaces (**e**), moment of time: $1 - \tau = 2.5 \times 10^5$ s; $2 - \tau = 5.0 \times 10^5$ s; $3 - \tau = 7.5 \times 10^5$ s.

As is shown in Figure 5a,b, the partial pressure of water vapor, as well as relative air humidity inside the material, decrease with time. The maximum values of these quantities are observed in the middle section of the material. In the direction of heat and mass exchange surfaces ($z = 0$ and $z = 0.1$ m), these quantities decrease to their values in the external medium. The material moisture content w_l changes in a similar way to w_l (Figure 5d).

As a result, at the initial time, the material and environment temperature are identical, while the material internal energy is spent on the evaporation process in the initial period of heat and mass transfer. Therefore, its temperature initially decreases and becomes lower than the initial value. Then, as the external medium temperature becomes higher than the material temperature, heat flows into the material from the outside. This heat is spent on the evaporation process and the gradual material heating. Its temperature rises over time (Figure 5c).

The distribution of evaporated vapor mass flows $J_{l\_v,dif}$ over the material thickness is shown in Figure 5e. As is shown in this figure, the evaporation process inside the material most intensely occurs in the areas near its surfaces. Over time, the maxima of curves $J_{l\_v,dif}(z)$ gradually move into the material.

In the second example, a material with the same initial parameters is placed in an air medium at 35 [°C]. The partial density of water vapor in the air medium is the same as in the first case: $\rho_{v,\infty} = 0.0104$ [kg/m^3]. Naturally, the relative humidity of the external air medium falls down to $\varphi = 0.26$.

The calculation results of the variation with time in temperature and moisture conditions of the capillary-porous material for these conditions are shown in Figure 6.

As can be seen from a comparison of Figure 6 with Figure 5, the behavior of the water vapor partial pressure with time, the relative air humidity and moisture content inside the material, is basically the same as in the previously considered case, when the initial material temperature was the same as external medium temperature. However, when the medium temperature outside the material is higher than the initial material temperature, its drying is much more intensive.

It has been proven by the moisture content degree in the material w_l for the same time intervals, analyzed in the first variant. If in the first variant at $\tau = 7.5 \times 10^5$ [s], the maximum moisture content in the middle part of the material, with its maximum, is $w_l = 35.65$ [kg/m^3] (Figure 5d), then in the second variant, the specified moisture content will be $w_l = 7.48$ [kg/m^3] (Figure 6d).

Unlike the first variant, the temperature inside the material changes with time. From the initial time, the porous material temperature starts to increase due to the initial temperature difference between the material and the external medium. The heat, entering the material from the outside, is spent on material heating and liquid evaporation inside the material.

The distribution of evaporated vapor mass flows $J_{l\_v,dif}$ over the material thickness is shown in Figure 6e.

As in the previously analyzed variant, the evaporation process inside the material in the initial period occurs most intensely in the areas near its surfaces. Over time, the maxima of curves $J_{l\_v,dif}(z)$ gradually move into the material and merge into one maximum in its middle.

3.2. Condensation Processes

The next example considers the case when the investigated porous material, which at the initial moment of time has a temperature of $t_0 = 35$ °C and a moisture content $w_{l,0} = 3.8$ [kg/m^3], is placed in an air environment with a temperature of $t_\infty = 20$ [°C] and a relative humidity of $\varphi_\infty = 0.6$ In this case, as in the previous cases, $\rho_{v,\infty} = 0.0104$ [kg/m^3]. Under such conditions, the process of moisture condensation begins on the surfaces of the sample, and then in its inner layers. The results of calculating the change in time of the temperature-humidity state of the sample under these conditions are shown in Figure 7.

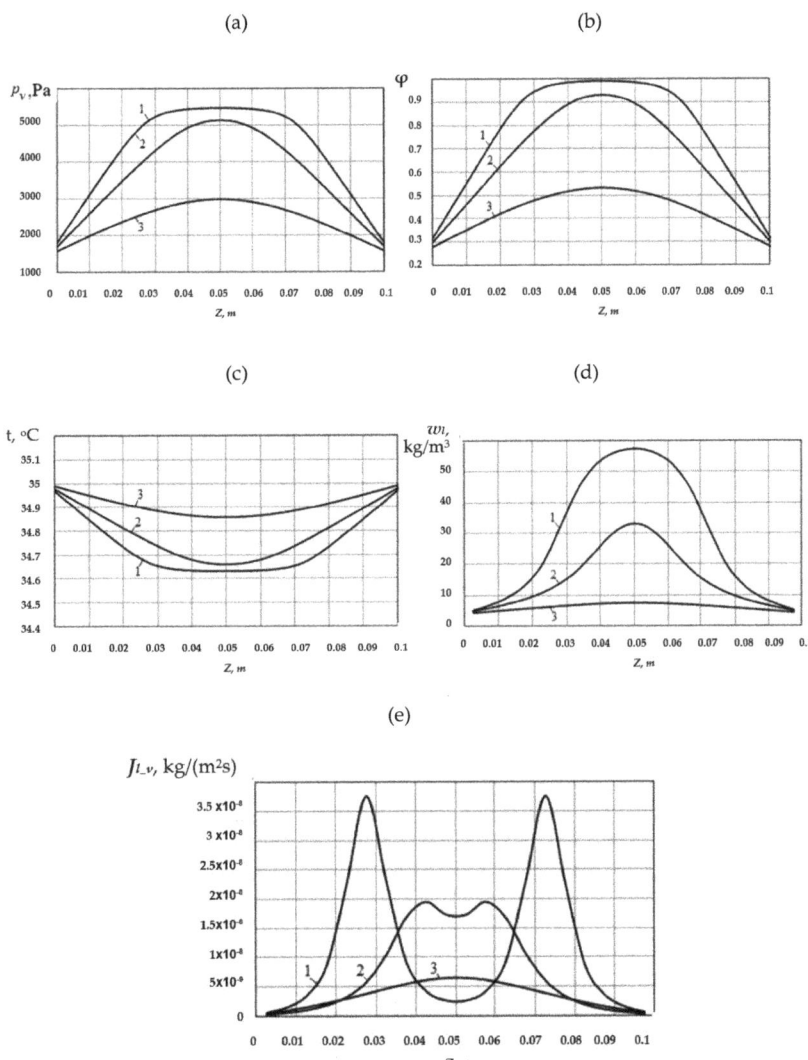

Figure 6. Distribution of the vapor partial pressure over the material thickness (**a**), relative air humidity (**b**), temperature (**c**), moisture content (**d**), and vapor flow density from the liquid column surfaces (**e**), moment of time: $1 - \tau = 2.5 \times 10^5$ [s]; $2 - \tau = 5.0 \times 10^5$ [s]; $3 - \tau = 7.5 \times 10^5$ [s]

As seen from Figure 7a,b, as a result of moisture condensation on the surfaces and inside the material, the partial pressure of water vapor, as well as the relative humidity of the air inside the material, increases over time. Over time, the moisture content of the porous material also increases (Figure 7e) and approaches the value of the maximum hygroscopic value corresponding to the humidity of the outside air. The temperature on the surfaces and inside the material gradually decreases (Figure 7c,d).

The modulus of the vapor flux density $J_{l\_v,dif}(z)$, which condenses inside the material (the flux itself is negative), has its maximum values at the surface in the initial period. Over time, the maximum value of the modulus of the vapor flow $J_{l\_v,dif}(z)$, which condenses inside the material, moves to the middle of the sample.

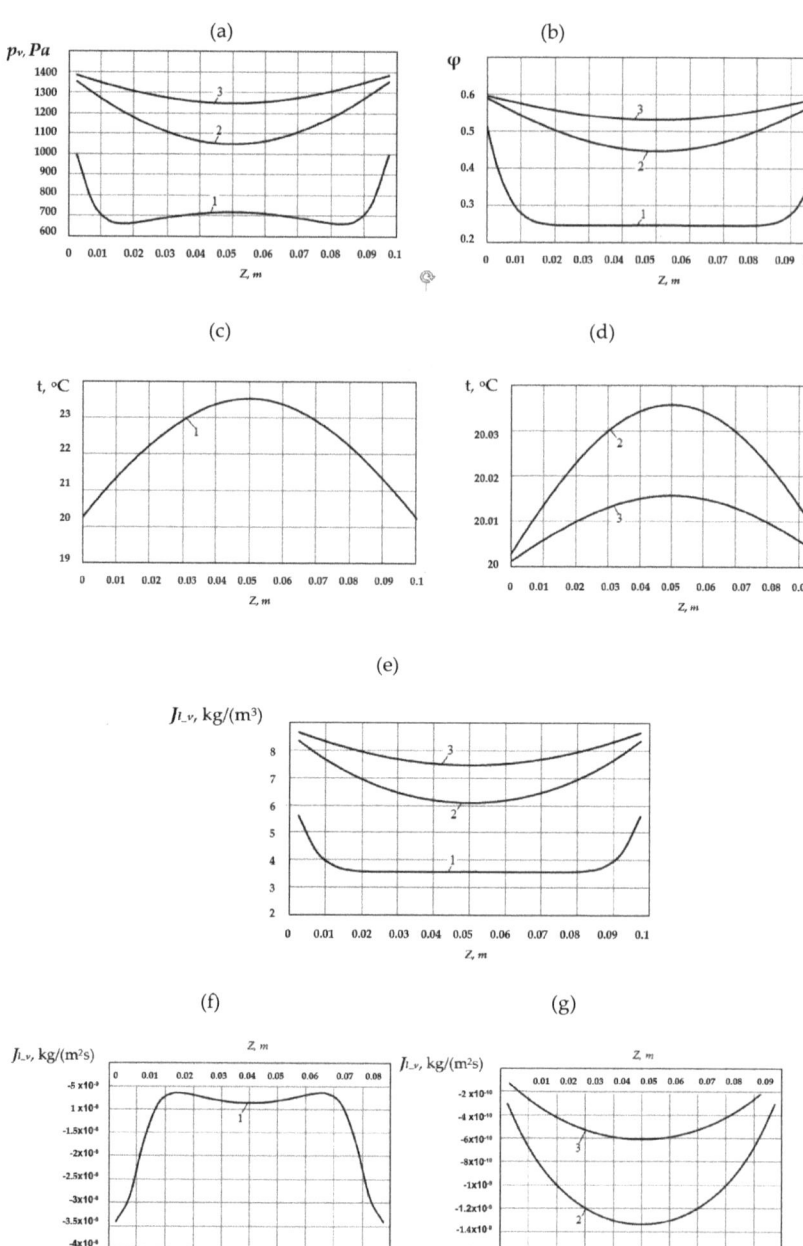

Figure 7. Distribution of the vapor partial pressure (**a**), relative air humidity (**b**), temperature (**c**, **d**), moisture content (**e**), and vapor flow density from the liquid column surfaces (**f**; **g**) over the material thickness and moment of time: $1 - \tau = 5 \times 10^3$ [s]; $2 - \tau = 2.5 \times 10^5$ [s]; $3 - \tau = 5.0 \times 10^5$ [s].

4. Discussion

As follows from the presented results, the analyzed network model of a wet capillary-porous material can be used to calculate the dynamics of changes in its temperature and moisture conditions.

This model makes it possible to calculate distribution over the thickness and change in time of the partial pressure of water vapor, temperature and liquid moisture content inside the material with the changes in temperature and moisture of the outside air.

According to the results of computational studies, evaporation (or condensation) inside the pores of a material with a change in external conditions occurs more intensively near its boundaries. Over time, the most intense areas of evaporation pass into the depth of the material. Note that the dynamics of temperature change are more intense than the dynamics of changes in humidity and moisture content.

For all considered cases, the times of establishment of thermodynamic equilibrium are rather long at more than 10 days (more than 7.5×10^5 s). Such dynamics logically correspond to the physics of the resulting effects.

The model is non-equilibrium, it is based on the differences in the parameters of the state of the vapor-gas and liquid phases in the micropores of the material. The difference in temperatures reached up to 1.5 [°C], in pressures-up to 5 [MPa], mainly due to the capillary pressure in the microchannel with a solid liquid. By the way, the capillary pressure was not specified by an analytical expression containing the surface tension, but the original tabular data [9] were used, taking into account the deviation of the microchannel from a strict cylindrical shape, for example, its possible cone shape.

The model is not devoid of limitations and incompleteness of taking into account the accompanying physical effects. It is applicable for a temperature range of no less than 0 [°C] and no more than 100 [°C]. At subzero temperatures, water freezes (or ice melts), and such a phase transformation is not taken into account. At temperatures above 100 [°C], it is necessary to complicate the model and take into account the effects of volumetric boiling of the liquid. The model also does not take into account the adhesion of vapor molecules on the surface of the solid material of the walls of the microchannel, does not take into account the possible film flow over the surface of the walls and does not take into account possible structural modes of liquid flow in the microchannel, such as slug, foam, dispersed and other flows.

Possible further studies of the proposed model are as follows. First of all, this could involve checking the model for: sensitivity to changes in fixed parameters and characteristics of a solid material (pore diameter, integral porosity index, its permeability); dependence of thermophysical characteristics on temperature and pressure; the structure of the filtration fluid flow; and other factors. It is of interest to make similar calculations for other materials, for example, thermal insulation. It is possible to develop a model for a different pore shape, for example, a spherical one.

It is extremely interesting to compare the model calculations with some data that were previously obtained by the authors in the experimental study of the dynamics of changes in the moisture content of a number of building and heat-insulating materials, depending on the humidity of the surrounding air.

In studying the condensation processes, calculations showed that in the first 5 min the specific mass flows of water vapor are very high—they reach 10^{-6} [kg/(m^2·s)] and higher. It can result in the formation of a continuous film of dropping liquid (water) on the material (brick) surface and complete filling of pores with water in a thin near-surface layer to a material depth of 0.5...1 [mm]. When the ambient temperature drops to sub-zero values (in degrees Celsius), this condition can result in the ice formation in near-surface micropores. A further decrease in temperature is accompanied by volumetric expansion of ice, leading to microdestructions of the material surface, leading to a loss of surface strength. Therefore, the modeling results of water vapor condensation can be applied to engineering calculations in the processes and technologies against the surface microdestruction in facade building structures.

5. Conclusions

The presented model includes a certain number of parameters, thermophysical properties and characteristics of a porous material. Some of them depend heavily on moisture content. It refers chiefly to capillary pressure and filtration coefficients. The computational model also describes the dependence of the material equilibrium moisture content on the air relative humidity (sorption curve). To provide solid results on the temperature-moisture condition of a porous material, based on the proposed calculation model, reliable data on the specified characteristics of materials are required. These characteristics for specific materials should be obtained from complex experimental studies using special laboratory facilities, which is a research problem to be solved. To derive required thermophysical characteristics of the studied material from the experimental data, we may use the proposed transfer model to solve inverse problems of heat and mass transfer.

It is also important to obtain reliable information on the structure of porous materials based on modern optical or electronic microscopy, using fluorescent substances that fill the pores.

The developed model can be effectively used in describing the processes of drying capillary-porous materials; in fact, from the problem involving this area of heat and mass transfer, the original problem statement arose. This model is probably not quite suitable for studying colloidal structures.

It is also advisable to thoroughly check the model (verification or validation) using other numerical modeling approaches, for example, using the LBM model or direct CFD-modeling.

Undoubtedly, such bifurcations of the further use of the model will require its corresponding correction, adjustment and, of course, time.

Author Contributions: B.B.: Conceptualization, Investigation, Writing—original draft; B.D.: Conceptualization, Investigation, Writing—original draft, Data analyses, Visualization; A.M.P.: Conceptualization, Investigation, Writing—editing. All authors have read and agreed to the published version of the manuscript.

Funding: This research was funded by grant number 025/RID/2018/19 "Regional Initiative of Excellence" in 2019–2022.

Data Availability Statement: The data presented in this study is available upon request from the respective author. The article presents a new mathematical model and calculation method that are available for use. A computer program may be available from the authors of the article.

Conflicts of Interest: The author declares no conflict of interest.

References

1. Whitaker, S. Coupled transport in multiphase systems: A theory of drying. *Adv. Heat Transf.* **1998**, *31*, 1–104.
2. Jiang, P.X.; Lu, X.C. Numerical simulation of fluid flow and convection heat transfer in sintered porous plate channels. *Int. J. Heat Mass Transf.* **2006**, *49*, 1685–1695. [CrossRef]
3. Kadem, S.; Lachemet, A.; Younsi, R.; Kocaefe, D. 3d-Transient modeling of heat and mass transfer during heat treatment of wood. *Int. Commun. Heat Mass Transf.* **2011**, *38*, 717–722. [CrossRef]
4. Giorgio, P.; Jianchao, C.; Zhien, Z.; Shimin, L. Advances in Modelling of Heat and Mass Transfer in Porous Materials. *Hindawi Adv. Mater. Sci. Eng.* **2019**. [CrossRef]
5. Junfeng, L.U.; Wen-qiang, L.U. Review: Heat and mass transfer in porous medium, mathematic/numerical models and research directions. *J. Int. J. Petrochem. Sci. Eng.* **2018**, *3*, 97–100.
6. Mirosław, S.; Michał, W.; Piotr, Ł.; Piotr, F.Ł. Analysis of Non-Equilibrium and Equilibrium Models of Heat and Moisture Transfer in a Wet Porous Building Material. *Energies* **2020**, *13*, 214. [CrossRef]
7. Defraeye, T. Advanced computational modelling for drying processes-a review. *Appl. Energy* **2014**, *131*, 323–344. [CrossRef]
8. Janetti, B.M.; Colombo, L.P.M.; Ochs, F.; Feist, W. Effect of evaporation cooling on drying capillary active building materials. *Energy Build.* **2018**, *166*, 550–560. [CrossRef]
9. Chu, S.-S.; Fan, T.-H.; Chang, W.-J. Modelling of coupled heat and moisture transfer in porous construction materials. *Math. Comput. Model.* **2009**, *50*, 1195–1204. [CrossRef]
10. Kubis, M.; Wisniewski, T.S.; Jaworski, M. Preliminary mathematical and numerical transient models of convective heating and drying of a brick. *MATEC Web. Conf.* **2018**, *240*, 01022.
11. Wasik, M.; Cieslikiewicz, Ł.; Łapka, P.; Furmanski, P.; Kubis, M.; Seredynski, M.; Pietrak, K.; Wisniewski, T.S.; Jaworski, M. Initial credibility analysis of a numerical model of heat and moisture transfer in porous building materials. *AIP Conf. Proc.* **2019**, *2078*, 020106.

12. Seredynski, M.; Wasik, M.; Łapka, P.; Furmanski, P.; Cieslikiewicz, Ł.; Pietrak, K.; Kubis, M.; Wisniewski, T.S.; Jaworski, M. Investigation of the equilibrium and non-equilibrium models of heat and moisture transport in a wet porous building material. *E3S Web. Conf.* **2019**, *128*, 06008. [CrossRef]
13. Allam, R.; Issaadi, N.; Belarbi, R.; El-Meligy, M.; Altahrany, A. Hygrothermal behavior for a clay brick wall. *Heat Mass Transf.* **2018**, *54*, 1579–1591. [CrossRef]
14. Karima, S.; M'barek, F.; Nabila, L.; M'hand, O.; Youb, K.B. Numerical Study of Heat and Mass Transfer during the Evaporative Drying of Porous Media. *MATEC Web. Conf.* **2020**, *307*, 01050.
15. Yingying, W.; Chao, M.; Yanfeng, L.; Dengjia, W.; Jiaping, L. Effect of moisture migration and phase change on effective thermal conductivity of porous building materials. *Int. J. Heat Mass Transfer.* **2018**, *125*, 330–342.
16. Verma, A.; Pitchumani, R. Fractal description of microstructures and properties of dynamically evolving porous media. *Int. Commun. Heat Mass Transfer.* **2017**, *81*, 51–55. [CrossRef]
17. Akulich, P.V. Heat and mass transfer in cylindrical porous bodies with account of nonstationarity of parameters of deepening evaporation. *Proc. Natl. Acad. Sci. Belarus Phys. Tech. Ser.* **2016**, *3*, 71–75.
18. Lin, Q.H.; Zouab, D.T.; Dongsheng, W.; Yanhui, F.; Xinxin, Z. Applied Thermal Engineering Inhomogeneity in pore size appreciably lowering thermal conductivity for porous thermal insulators. *Appl. Therm. Eng.* **2018**, *130*, 1004–1011.
19. Faeez, A.; Arman, R.; Evangelos, T.; Marc, P.; Abdolreza, K. From micro-scale to macro-scale modeling of solute transport in drying capillary porous media. *Int. J. Heat Mass Transfer.* **2021**, *165*, 120722.
20. Zilong, D.; Xiangdong, L.; Yongping, H.; Chengbin, Z.; Yongping, C. Heat Conduction in Porous Media Characterized by Fractal Geometry. *Energies* **2017**, *10*, 1230.
21. El Moumen, A.; Kanit, T.; Imad, A.; El Minor, H. Computational thermal conductivity in porous materials using homogenization techniques: Numerical and statistical approaches. *Comput. Mater. Sci.* **2015**, *97*, 148–158. [CrossRef]
22. Yuan, Y.; Zhao, Z.; Nie, J.; Xu, Y. Pore Network Analysis of Zone Model for Porous Media Drying. *Math. Probl. Eng.* **2014**, *2014*. [CrossRef]
23. Andreas, G.; Yiotis, I.N.; Tsimpanogiannis, A.K.; Stubos, Y.C.Y. Pore-network study of the characteristic periods in the drying of porous materials. *J. Colloid Interface Sci.* **2006**, *297*, 738–748.
24. Surasani, V.K.; Metzger, T.; Tsotsas, E. Drying Simulations of Various 3D Pore Structures by a Nonisothermal Pore Network Model. *Dry. Technol.* **2010**, *28*, 615–623. [CrossRef]
25. Jan, C.; Filip, D.; Geert, H. A Multiscale Network Model for Simulating Moisture Transfer Properties of Porous Media. *Transp. Porous Media* **1999**, *35*, 67–88.
26. Raoof, A.; Nick, H.M.; Hassanizadeh, S.M.; Spiers, C.J. Pore Flow: A complex pore-network model for simulation of reactive transport in variably saturated porous media. *Comput. Geosci.* **2013**, *61*, 160–174. [CrossRef]
27. Alireza, A.; Moghaddam, M.P.; Evangelos, T.; Abdolreza, K. Evaporation in Capillary Porous Media at the Perfect Piston-Like Invasion Limit: Evidence of Nonlocal Equilibrium Effects. *Water Resour. Res.* **2017**, *53*, 0433–10449.
28. Nicole, V.; Thomas, M.; Evangelos, T. On the Influence of Temperature Gradients on Drying of Pore Networks. In Proceedings of the European Drying Conference-EuroDrying'2011, Palma, Spain, 26–28 October 2011.
29. Nicole, V.; Wang, Y.J.; Abdolreza, K.; Evangelos, T.; Marc, P. Drying with Formation of Capillary Rings in a Model Porous Medium. Transport in Porous Media. *Springer Verlag* **2015**, *110*, 197–223.
30. Moghaddam, A.; Attari, K.A.; Tsotsas, E.; Prat, M. A pore network study of evaporation from the surface of a drying non-hygroscopic porous medium. *AIChE J.* **2018**, *64*, 1435–1447. [CrossRef]
31. Faeez, A.; Marouane, T.; Marc, P.; Evangelos, T.; Abdolreza, K. Non-local equilibrium continuum modeling of partially saturated drying porous media: Comparison with pore network simulations. *Chem. Eng. Sci.* **2020**, *228*, 115957.
32. Xiang, L.; Abdolreza, K.; Evangelos, T. Transport parameters of macroscopic continuum model determined from discrete pore network simulations of drying porous media. *Chem. Eng. Sci.* **2020**, 115723. [CrossRef]
33. Xiang, L.; Abdolreza, K.; Hadi, A.; Evangelos, T. The Brooks and Corey Capillary Pressure Model Revisited from Pore Network Simulations of Capillarity-Controlled Invasion Percolation Process. *Processes* **2020**, *8*, 10.
34. Bultreys, T.; De Boever, W.; Cnudde, V. Imaging and image-based fluid transport modeling at the pore scale in geological materials: A practical introduction to the current state-of-the-art. *Earth Sci. Rev.* **2016**, *155*, 93–128. [CrossRef]
35. Agaesse, T.; Lambrac, A.; Büchi, F.N.; Pauchet, J.; Prat, M. Validation of pore network simulations of ex-situ water distributions in a gas diffusion layer of proton exchange membrane fuel cells with X-ray tomographic images. *J. Power Sources* **2016**, *331*, 462–474. [CrossRef]

Article

Adsorptive and Electrochemical Properties of Carbon Nanotubes, Activated Carbon, and Graphene Oxide with Relatively Similar Specific Surface Area

Krzysztof Kuśmierek [1], Andrzej Świątkowski [1], Katarzyna Skrzypczyńska [2] and Lidia Dąbek [3],*

[1] Institute of Chemistry, Military University of Technology, 00-908 Warsaw, Poland; krzysztof.kusmierek@wat.edu.pl (K.K.); andrzej.swiatkowski@wat.edu.pl (A.Ś.)
[2] Łukasiewicz—Industrial Chemistry Research Institute, 01-793 Warsaw, Poland; katarzyna.skrzypczynska@ichp.pl
[3] Faculty of Environmental, Geomatic and Energy Engineering, Kielce University of Technology, 25-314 Kielce, Poland
* Correspondence: ldabek@tu.kielce.pl; Tel.: +48-41-34-24-869

Citation: Kuśmierek, K.; Świątkowski, A.; Skrzypczyńska, K.; Dąbek, L. Adsorptive and Electrochemical Properties of Carbon Nanotubes, Activated Carbon, and Graphene Oxide with Relatively Similar Specific Surface Area. *Materials* **2021**, *14*, 496. https://doi.org/10.3390/ma14030496

Academic Editor: Alexander A. Lebedev
Received: 30 December 2020
Accepted: 18 January 2021
Published: 21 January 2021

Publisher's Note: MDPI stays neutral with regard to jurisdictional claims in published maps and institutional affiliations.

Copyright: © 2021 by the authors. Licensee MDPI, Basel, Switzerland. This article is an open access article distributed under the terms and conditions of the Creative Commons Attribution (CC BY) license (https://creativecommons.org/licenses/by/4.0/).

Abstract: Three carbon materials with a highly diversified structure and at the same time much less different porosity were selected for the study: single-walled carbon nanotubes, heat-treated activated carbon, and reduced graphene oxide. These materials were used for the adsorption of 2,4-D herbicide from aqueous solutions and in its electroanalytical determination. Both the detection of this type of contamination and its removal from the water are important environmental issues. It is important to identify which properties of carbon materials play a significant role. The specific surface area is the major factor. On the other hand, the presence of oxygen bound to the carbon surface in the case of contact with an organochlorine compound had a negative effect. The observed regularities concerned both adsorption and electroanalysis with the use of the carbon materials applied.

Keywords: single-walled carbon nanotubes; activated carbon; reduced graphene oxide; 2,4-D; adsorption; electroanalysis

1. Introduction

The widespread use of chlorophenoxy herbicides in agriculture is an important source of soil and water contamination [1]. One of the most important herbicides, 2,4-dichlorophenoxyacetic acid (2,4-D), is used in several countries to control weeds. Even over 70 years after its introduction, 2,4-D continues to be the most common and widely used herbicide worldwide [2,3]. It is directly applied onto soil or sprayed over crop fields and, from there, often reaches surface waters and sediments. The choice of 2,4-D as the test substance was dictated by the fact that it belongs to a group of organochlorine pollutants to surface waters and groundwater which are characterized by high harmfulness of living organisms. The purpose of this work was to evaluate the adsorption potential of chosen carbon materials with a strongly differentiated structure for 2,4-D as the target water contaminant. So far, in research on the effectiveness of use as adsorbents against herbicides or phenol derivatives, activated carbons differing in porosity or surface chemistry were most often used [4–7]. Carbon materials with a different internal structure were used relatively rarely in such studies [8–10]. Moreover, many studies used carbon materials of various types with very different porosities (S_{BET}) [8–10], which made it practically impossible to assess the influence of their internal structure differences on the adsorption efficiency.

Carbon materials, especially those classified into the graphite family, have a variety of textures (nanotexture and microtexture) and structures [11,12].

The graphite family refers to carbon materials that are similar to graphite in the sp² hybridization of carbon atoms and the presence of several hexagonal carbon layers in the structure. There are numerous members of this family. For example, activated carbon that is formed by subjecting turbostratic carbon to a chemical reaction (activation) that consumed a part of the carbon. The activation results in surface pores, which cause the specific surface area to be high, as needed for fluid purification through the adsorption of the molecules to be removed.

The graphite family also includes graphite oxide, which is a covalent form of intercalated graphite. The conversion of graphite to graphite oxide involves its oxidation. The separation of the layers is a step in one of the methods of making graphene. The further reduction results in the removal of the oxygen and forming graphene oxide and, at the end, the reduced material known as reduced graphene oxide (rGO). The rGO sheets are usually considered as one kind of chemically derived graphene.

Carbon nanotubes (CNTs) are cylindrical molecules that consist of rolled-up sheets of single-layer carbon atoms (graphene). Among them, there are single-walled nanotubes (SWCNTs) with a diameter even less than 1 nm and multi-walled nanotubes (MWCNTs), consisting of several concentrically inter-linked nanotubes (diameters more than 100 nm). Like their building block graphene, CNTs are chemically bonded with sp^2 bonds.

Carbon materials of a completely different type of structure were selected and used for the studies. The only common feature for them was a similar value of the specific surface area. The materials selected were commercial single-walled carbon nanotubes and reduced graphene oxide, as well as demineralized and high temperature, heated activated carbon to reduce its specific surface area.

In our work, apart from the adsorption properties of carbon materials, their electroanalytical usefulness was also tested. The electrochemical properties of carbon (graphite) paste electrodes (CPEs) modified by the addition of these carbon materials and their application in electroanalysis of 2,4-D were investigated.

2. Materials and Methods

2.1. Reagents and Materials

The graphite powder (45 µm), spectroscopic grade paraffin oil, as well as the 98% 2,4-dichlorophenoxyacetic acid (2,4-D), were purchased from Sigma Aldrich (St. Louis, MO, USA). Commercial, high-purity, open, single-walled carbon nanotubes (SWCNTs), from Nanostructured & Amorphous Materials, Inc. (Houston, TX, USA), were chosen for investigation. The second used carbonaceous nanomaterial was commercially reduced graphene oxide (rGO) received from Advanced Graphene Products Sp. z o.o. (Nowy Kisielin, Zielona Góra, Poland). The final carbon material used for comparison was extruded commercial activated carbon R3ex (Norit) demineralized with concentrated HF and HCl acids and next annealed in argon at 1800 °C (AC1800) [13]. All three carbon materials were selected in such a way that their specific surface area differs not too much. At the same time, their structure was extremely diverse.

2.2. Adsorbents Characterization

2.2.1. Adsorption–Desorption N_2 Isotherms

The porous structure of the carbon materials used for investigations was characterized using nitrogen adsorption–desorption isotherms at 77.4 K (ASAP 2020, Micromeritics, Norcross, GA, USA). On this basis, the main parameters characterizing the porosity of the chosen carbon materials, the specific surface area (S_{BET}), micro- (V_{mi}), and mesopore (V_{me}) volumes were calculated.

2.2.2. Microscopic Studies

Measurements of the surface-bonded oxygen content were carried out using a scanning microscope (Philips XL30/LaB6, Amsterdam, Netherlands) coupled with an energy dispersive X-ray spectrometer (DX4i/EDAX device). SEM images were also determined. High-resolution transmission electron microscopy (HRTEM) images for SWCNT were taken using a transmission electron microscope F20X-TWIN (FEI-Tecnai) operated at 200 kV [14].

2.2.3. Raman Spectra

The characterization of the tested carbon materials was done by Raman spectroscopy using a Renishaw inVia Raman Microscope (Wotton-under-Edge, Gloucestershire, UK) with an exciting wavelength of 514.5 nm.

2.3. Adsorption from Aqueous Solutions

Batch experiments were carried out to examine the adsorption properties of the chosen carbon materials. All of the adsorption experiments were conducted at a room temperature of 25 °C in glassy Erlenmeyer flasks contacting different initial solutions of 2,4-D (0.02 L) with a given amount of carbon material (0.01 g).

The adsorption isotherms for the 2,4-D were constructed from solutions with an initial concentration ranging between 0.2 and 1 mmol/L. The flasks were then placed in the laboratory shaker and agitated at 100 rpm for 24 h. After this time, the samples were filtered through filter paper and analyzed for the 2,4-D content. The amount of the herbicide adsorbed per unit mass of adsorbent at equilibrium, q_e (mmol/g), was calculated from the following formula:

$$q_e = \frac{(C_0 - C_e)V}{m} \quad (1)$$

where C_0 and C_e (mmol/L) are 2,4-D concentrations at initial and final steps, respectively, m (g) is the mass of the adsorbent added, and V (L) is the volume of the solution.

The kinetic studies were conducted for an initial 2,4-D concentration of 0.5 mmol/L. The flasks were agitated at 100 rpm. Samples were taken at different preset contact time intervals, filtered to prevent the presence of adsorbent in the samples, and analyzed spectrophotometrically. The amount of 2,4-D adsorbed at time t, q_t (mmol/g), was evaluated by applying the following formula:

$$q_t = \frac{(C_0 - C_t)V}{m} \quad (2)$$

where C_t (mmol/L) is the 2,4-D concentration at time t.

The quantification of the 2,4-D presents in aqueous solution was made using UV–Vis spectrophotometry (Carry 3E, Varian, Palo Alto, CA, USA) at the wavelength of 278 nm. The calibration curve was constructed (0.02–0.8 mmol/L) by plotting absorbances vs. 2,4-D concentrations (y = 0.956x + 0.035; R^2 = 0.999).

2.4. Voltammetry

All the voltammetric measurements were conducted with an AutoLab PGSTAT 20 (Eco Chemie, Utrecht, Netherlands) potentiostat. For all electrochemical measurements, the conventional three-electrode system was used: (1) a modified carbon paste electrode, (2) a Pt wire, and (3) a saturated calomel electrode, which functioned as the working electrode, the auxiliary electrode, and the reference electrode, respectively. Differential pulse voltammetry (DPV) was used in this study. The measurements were carried out in a homemade 40 mL glass cell. Voltammograms were registered from 0 to 2.0 V at a sweep rate of 50 mV/s. The pulse height and width were set as 50 mV and 50 ms, respectively, and the sampling time was 50 ms. A Teflon holder with a hole at one end for filling the carbon paste served as the electrode body. Electrical contact was made with a stainless-steel rod through the center of the holder. Modified CPEs were prepared by thoroughly mixing 2.5, 5, or 10 wt.% of modifying carbon material and graphite powder with the subsequent addition of mineral oil. All ingredients were placed in an agate mortar, crushed with a pestle and the mixture was kept at room temperature for 3 days. The prepared paste was then packed into the hole of the electrode body and smoothed onto a paper.

3. Results and Discussion

3.1. Physicochemical Characterization of the Carbon Materials

The low-temperature nitrogen adsorption–desorption isotherms on the carbon materials are presented in Figure 1. The specific surface areas (S_{BET}), as well as the micropore (V_{mi}) and mesopore (V_{me}) volumes, were calculated from the N_2 adsorption isotherms, and the results are listed in Table 1. The S_{BET} was calculated by the BET method, while the V_{mi} and V_{me} volumes were calculated using the t-plot method.

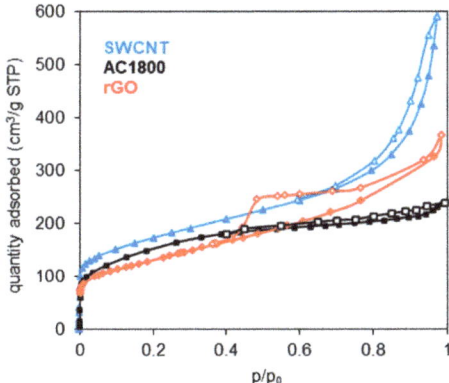

Figure 1. Nitrogen adsorption–desorption isotherms (measured at 77 K) of single-walled carbon nanotubes (SWCNTs), AC1800, and rGO.

Table 1. Physicochemical properties of the carbon materials tested.

Adsorbent	S_{BET} (m^2/g)	V_{mi} (cm^3/g)	V_{me} (cm^3/g)	$V_{mi}/V_{mi} + V_{me}$	Oxygen Surface Content, (% wt.)
SWCNT	597	0.267	0.314	0.460	4.9
AC1800	554	0.239	0.133	0.642	1.4
rGO	512	0.220	0.272	0.447	17.1

Table 1 shows that the specific surface areas of the tested materials are not very diverse (the maximum difference between them is 85 m^2/g). A similar relationship can be noticed in the case of the micropore volume, while there are significant differences in the mesopore volume—the AC1800 exhibits the lowest V_{me} as a result of its high-temperature treatment [7,13].

The results of the performed HRTEM analysis for SWCNT [14] are presented in Figure 2a–c. As can be observed, CNTs are long and partially opened. On the external walls, some amorphous carbon forming the debris is observed. The images (Figure 2) show a large variation in the surface morphology of the tested carbon materials, resulting from their internal structure.

Figure 2. The TEM (SWCNT) [14] and SEM images of the carbon materials. (**a**) SWCNT; (**b**) AC1800; (**c**) rGO.

Figure 3 shows the Raman spectra obtained for the tested carbon materials. The peaks at about 1350 and 1580 cm^{-1} correspond to D and G bands, respectively. The D-band represents the A_{1g} vibration mode caused by the disordered structure of the carbon materials, whereas the G-band corresponds to the E_{2g} vibration mode in the graphitic lattice of carbon materials [15].

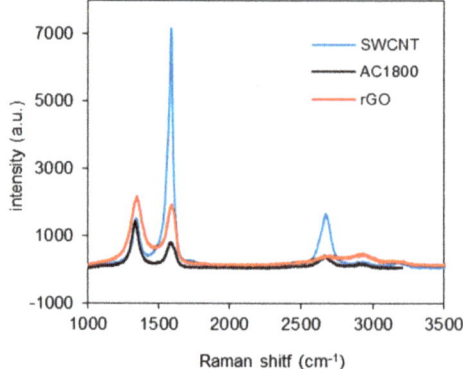

Figure 3. The Raman spectra of the SWCNT, AC1800, and rGO.

For the carbon materials used in this research, the D-band was located at 1348, 1350, and 1349 cm^{-1} for SWCNT, AC1800, and rGO, respectively, and the G-band was at 1573, 1581, and 1588 cm^{-1}, respectively. Using the ratio of peak intensities I_D/I_G, one can use Raman spectra to characterize the disorder level in carbon materials. The appropriate values were 0.031, 2.21, and 1.30 for SWCNT, AC1800, and rGO, respectively. As one can see, the carbon materials tested are highly differentiated in terms of internal structure arrangement. SWCNT shows the highest difference in intensities of the D and G bands [14,16]. The opposite situation can be observed for the heat-treated activated carbon AC1800. The heating temperature (1800 °C), although significantly reducing its pore volume, is too low for graphitization [13].

The relative intensity ratio of both peaks (I_D/I_G) for the rGO sample is a measure of disorder degree and is inversely proportional to the average size and number of the sp^2 clusters [15,17].

3.2. Adsorption Study

The adsorption of the 2,4-D from aqueous solutions onto single-walled carbon nanotubes, heat-treated activated carbon, and reduced graphene oxide was studied by means of the adsorption kinetics and the construction of adsorption isotherms.

Adsorption kinetics of 2,4-D from the water on the carbonaceous materials is shown in Figure 4. It was observed that the adsorption rapidly increased in the first steps of the process and reached the adsorption equilibrium within about 1–2 h.

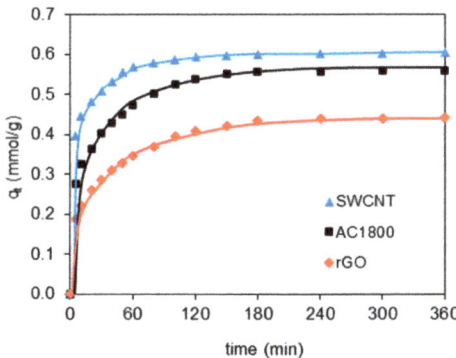

Figure 4. The effect of contact time on 2,4-D adsorption onto SWCNT, AC1800, and rGO.

To further analyze the adsorption kinetics, the data from Figure 4 were fitted by pseudo-first-order (PFO) (Equation (3)) and pseudo-second-order (PSO) (Equation (4)) kinetic models:

$$\log(q_e - q_t) = \log q_e - \frac{k_1}{2.303} t \tag{3}$$

$$\frac{t}{q_t} = \frac{1}{k_2 q_e^2} + \frac{1}{q_e} t \tag{4}$$

where k_1 and k_2 are the rate constants of PFO (1/min) and PSO adsorption (g/mmol·min), respectively.

The values of k_1 and k_2 were calculated from the slope and intercept of the plots of $\log(q_e - q_t)$ versus t and t/q_t versus t, respectively, and are given in Table 2.

Table 2. The pseudo-first- and pseudo-second-order rate constants for adsorption of 2,4-D on the carbonaceous materials.

Adsorbent		Pseudo-First-Order			Pseudo-Second-Order		
	$q_{e(EXP)}$ (mmol/g)	k_1 (1/min)	R^2	$q_{e(CAL)}$ (mmol/g)	k_2 (g/mmol·min)	R^2	$q_{e(CAL)}$ (mmol/g)
SWCNT	0.604	0.022	0.935	0.423	0.049	0.999	0.612
AC1800	0.559	0.019	0.975	0.323	0.016	0.999	0.579
rGO	0.441	0.021	0.992	0.305	0.038	0.999	0.462

The R^2 values for the PSO kinetic model are equal or greater than 0.999 for adsorption of the herbicide on all of the carbonaceous materials. Furthermore, a better agreement between the experimental ($q_{e(EXP)}$) and calculated ($q_{e(CAL)}$) values of equilibrium adsorption capacity was observed for the PSO kinetic model than for the PFO. This suggests that the adsorption of the 2,4-D on the adsorbents follows the pseudo-second-order kinetic model.

The adsorption results revealed that the 2,4-D was adsorbed faster on the SWCNT than on the rGO and that the microporous material of heat-treated activated carbon led to a much longer time to reach the adsorption equilibrium. The values of the k_2 for 2,4-D followed the sequence: AC1800 < rGO < SWCNT. This order in the rate of adsorption can be explained by the different porous structures of the materials—the content of mesopores, which play the role of transporting arteries. The adsorption rate increased with an increase in the content of mesopores in the total porous structure of the carbon materials.

Adsorption isotherms of the 2,4-D on the SWCNT, AC1800, and rGO are presented in Figure 5. To understand the adsorption isotherm, the data from Figure 5 were fitted by the Langmuir (Equation (5)) and Freundlich (Equation (6)) models with the following nonlinear forms:

$$q_e = \frac{q_m b C_e}{1 + b C_e} \quad (5)$$

$$q_e = K_F C_e^{1/n} \quad (6)$$

where q_m (mmol/g) is the maximum adsorption capacity, b (L/mmol) is the Langmuir parameter, while the K_F ((mmol/g)·(L/mmol)$^{1/n}$) and n are the Freundlich constants.

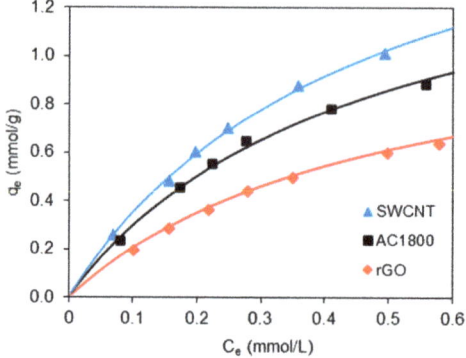

Figure 5. Adsorption isotherms of 2,4-D on SWCNT, AC1800, and rGO.

The suitability of these models was verified based on the correlation coefficient R^2 and standard deviation equation Δq (%) as:

$$\Delta q = 100 \times \sqrt{\frac{\Sigma\left[(q_{exp} - q_{cal})/q_{exp}\right]^2}{N-1}} \quad (7)$$

The results (Table 3) show that the Langmuir and Freundlich isotherm models fitted reasonably well the adsorption data for all of the adsorbents. However, the Langmuir model fitting was slightly better for experimental data due to the higher R^2 and lower Δq values. This suggested the monolayer and homogeneous adsorption of 2,4-D onto the SWCNT, AC1800, and rGO surface.

Table 3. The Langmuir and Freundlich isotherm equation parameters for adsorption of 2,4-D on the carbon materials.

Adsorption Model	Parameter	SWCNT	AC1800	rGO
Langmuir	q_m (mmol/g)	2.001	1.652	1.222
	b (L/mmol)	2.127	2.168	2.030
	R^2	0.991	0.993	0.994
	Δq (%)	2.139	2.888	4.244
Freundlich	K_F ((mmol/g)·(L/mmol)$^{1/n}$)	1.806	1.462	0.934
	n	1.403	1.439	1.540
	R^2	0.977	0.975	0.972
	Δq (%)	5.229	7.462	5.231

Equilibrium adsorption experiments revealed that the SWCNT had the highest 2,4-D adsorption capacity, compared to the AC1800 and rGO adsorption capacities. These findings can be attributed to the SWCNT's high specific surface area and micropore volume values that facilitate the efficient adsorption of 2,4-D. Both, the Langmuir and Freundlich parameters (q_m and K_F) follow the order of the BET specific surface areas (rGO < AC1800 < SWCNT). After converting the adsorption per 1 m^2 of surface area, a relatively small difference between SWCNT and AC1800 (0.0033 and 0.0030 mmol/m^2, respectively) can be observed, and in the case of rGO, this value (0.0024 mmol/m^2) was much lower than for the two previous materials.

The worst adsorption capacity of the rGO can be explained by taking into account one more factor besides S_{BET}, namely, the surface chemistry, and more specifically the amount of oxygen bound to the carbon surface. While in the case of SWCNT and AC1800, it was relatively small and comparable, in the case of rGO, it was very large (several times greater). The presence of the acidic surface oxygen groups decreases the adsorption efficiency. This phenomenon is associated with the hydration of polar carboxyl groups leading to the creation of water clusters, which can block active sites on the adsorbent surface and reduce its availability for adsorbent molecules [6,18]. In many studies, it was found that the presence of oxygen on the surface of carbon materials reduces the adsorption of organic compounds of similar structure to 2,4-D, e.g., 2-(4-chloro-2-methylphenoxy)acetic acid (MCPA), 2-(4-chloro-2-methylphenoxy)propanoic acid (MCPP), and 4-(4-chloro-2-methylphenoxy) butanoic acid (MCPB) [6].

Comparisons of 2,4-D monolayer adsorption capacity of the SWCNT, AC1800, and rGO with those of other adsorbents reported in previous studies are given in Table 4. The adsorption capacities reported in this study are high compared to other carbonaceous materials.

Table 4. Comparison of the 2,4-D adsorption on various adsorbents.

Adsorbent	S_{BET} (m^2/g)	Langmuir Adsorption Capacity, q_m (mg/g)	Ref.
SWCNT	597	442.3	this paper
AC1800	554	365.1	this paper
rGO	512	270.1	this paper
groundnut shell char	43	3.02	[19]
CB-C carbon black	97	68.6	[20]
CB-V carbon black	227	72.2	[20]
AC from sugarcane bagasse	507	153.9	[21]
AC from groundnut shell	709	250.0	[19]
commercial AC F-400	800	137.7	[22]
commercial AC SX-2	885	180.4	[23]
AC from coconut shell	991	233.0	[21]
commercial AC F-300	965	191.2	[23]
AC from data stones	763	238.0	[24]
AC from corncob	1274	300.0	[25]

3.3. Electroanalytical Research

The first stage of DPV studies was the evaluation of the effect of the accumulation time on the peak current of the oxidation of 2,4-D in a solution (0.5 mmol/L) for the CPEs modified with all the tested carbon materials. The current intensity was found to increase with the accumulation time until 7 min, after which it maintained a constant value (Figure 6a). In further studies, an accumulation time of 7 min was used.

Voltammograms for the CPE as an example with 10% content of carbon modifiers are shown in Figure 6b. From all the DPV curves (without as well as with 2.5%, 5%, and 10% modifiers content), the peak currents and the peak potentials were determined.

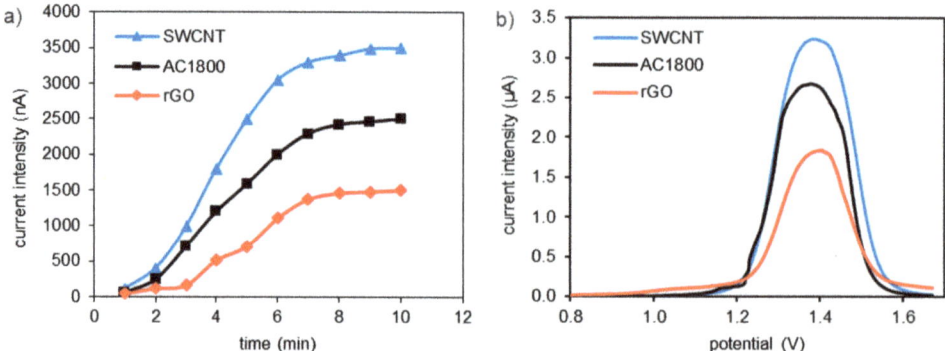

Figure 6. Effect of accumulation time on peak current for 0.5 mmol/L 2,4-D (**a**); differential pulse voltammetry (DPV) registered for 0.5 mmol/L 2,4-D solutions using carbon paste electrodes (CPEs) containing 10% by mass of the tested materials (**b**).

Similar to the dependence of the amount of adsorbed 2,4-D (q_m) on the value of the specific surface area (S_{BET}) of the carbon material, changes of the peak currents can be observed, recorded during the voltammetric measurements with the use of a carbon paste electrode modified by the addition of SWCNT, AC1800 or rGO. The values of the peak currents decrease with the decrease in the specific surface area of the CPE modifiers for different amounts of their additions and different concentrations of 2,4-D solutions. Similar relationships have already been described in the literature [7,8]. They have even been shown to be linear [8]. In our case, the decreasing relationship is also close to linear, but the peak current for CPE modified by the addition of rGO shows too much decrease.

For example, for the concentration of the 2,4-D solution 0.5 mmol/g and the content of the CPE modifiers 10% wt., the peak intensity (µA)/S_{BET} (m^2/g) ratios for SWCNT and AC1800 are 0.00552 and 0.00538, respectively, while for rGO it is only 0.00431. This too low value for rGO as a CPE modifier results, as can be considered, from the high oxygen content in this carbon material.

The differential pulse voltammetry was applied to determine the concentration of 2,4-D using the modified CPEs in the range of 0.002–0.5 mmol/L. Calibration curves were fitted by linear regression using the peak current versus concentration and the detection (LOD) and quantitation (LOQ) limits of the CPEs were calculated using the following formulas:

$$LOD = \frac{3\sigma}{a} \quad (8)$$

$$LOQ = \frac{6\sigma}{a} \quad (9)$$

where a is a slope of the calibration curve (y = ax + b), and σ is a standard deviation of the blank signal.

The linear regression equations, as well as the respective correlation coefficients, are presented in Table 5.

The sensitivity of all modified electrodes was much greater than that of the unmodified (graphite) electrode. Moreover, the sensitivity of the methods was correlated with the amount and type of modifier. The LOD decreased with the increase of the amount of the modifier content from 2.5 to 10% as well as with the S_{BET} of the carbon materials used. The best sensitivity (0.468 µmol/L) was observed for the CPE modified with SWCNT (10%). The sensitivity of these methods is comparable to other electrochemical methods for 2,4-D determination described elsewhere. The LOD was found to be 0.08 µmol/L for graphite-polyurethane electrode [26], 0.2 µmol/L for R3ex activated carbon modified CPE [8], 0.23 µmol/L for mercury electrode [27], 0.400 µmol/L for silica-gel modified CPE [28], 0.7 µmol/L for Carboxen 1000 carbon molecular sieve modified CPE [8], 0.83 µmol/L for electrochemical sensor based on molecularly imprinted polypyrrole membranes [29], 0.98 μmol/L for F-300 activated carbon modified CPE [23], 1.14 μmol/L for SX-2 activated carbon modified CPE [23], 3.15 μmol/L for bismuth film modified screen-printed carbon electrode [30], and 3.4 μmol/L for Carbopack B carbon black modified CPE [8].

Table 5. Linearity results for the modified carbon paste electrode.

Electrode	Linear Regression Equation y = ax + b	R^2	LOD (μmol/L)	LOQ (μmol/L)
graphite CPE	y = 0.054x + 0.006	0.992	50	100
SWCNT (2.5%) CPE	y = 3.484x + 0.076	0.999	0.775	1.549
SWCNT (5.0%) CPE	y = 4.931x + 0.109	0.997	0.555	1.095
SWCNT (10%) CPE	y = 5.760x + 0.138	0.998	0.468	0.937
AC1800 (2.5%) CPE	y = 3.111x + 0.015	0.992	0.868	1.736
AC1800 (5.0%) CPE	y = 4.833x + 0.089	0.994	0.561	1.117
AC1800 (10%) CPE	y = 5.378x + 0.171	0.995	0.502	1.004
rGO (2.5%) CPE	y = 2.480x − 0.029	0.997	1.089	2.177
rGO (5.0%) CPE	y = 3.631x + 0.074	0.995	0.744	1.487
rGO (10%) CPE	y = 4.061x + 0.161	0.996	0.673	1.330

4. Conclusions

The conducted research has shown that the adsorption and electrochemical properties of carbon materials in solutions of aromatic organochlorine compounds depend on, to a large extent, their porosity. In the case of the presence of a larger amount of oxygen bonded with their surface than usual (only a few % by weight), a worsening of their adsorption and electrochemical properties can be observed. The general conclusions are as follows:

- The higher the specific surface, the better the adsorption, as well as electroanalytical properties, of carbon materials;
- Suitability of carbon materials for adsorption and electroanalysis seems to be correlated;
- The internal structure of carbon materials affects their surface characteristics;
- In the case of organochlorine compounds, the presence of oxygen on the carbon surface reduces their adsorption.

Author Contributions: K.K.—conceptualization, investigation, writing—original draft preparation, visualization, writing—review and editing; A.Ś.—conceptualization, investigation, writing—original draft preparation, writing—review and editing, supervision; K.S.—investigation, visualization; L.D.—investigation, writing—review and editing. All authors have read and agreed to the published version of the manuscript.

Funding: The project was funded by the program of the Minister of Science and Higher Education entitled: "Regional Initiative of Excellence" in 2019–2022, project number 025/RID/2018/19, financing amount PLN 12,000,000.

Data Availability Statement: Data sharing is not applicable to this article.

Conflicts of Interest: The authors declare no conflict of interest.

References

1. Lagana, A.; Bacaloni, A.; de Leva, I.; Faberi, A.; Fago, G.; Marino, A. Occurrence and determination of herbicides and their major transformation products in environmental waters. *Anal. Chim. Acta* **2002**, *462*, 187–198. [CrossRef]
2. Moszczyński, W.; Białek, A. Ecological production technology of phenoxyacetic herbicides MCPA and 2,4-D in the highest world standard. In *Herbicides—Properties, Synthesis and Control of Weeds Chapter*; Hasaneen, M.N., Ed.; IntechOpen: Rijeka, Croatia, 2012; pp. 349–362.
3. Białek, A.; Moszczyński, W. Technological aspects of the synthesis of 2,4-dichlorophenol. *Pol. J. Chem. Technol.* **2009**, *11*, 21–30. [CrossRef]

4. Salame, I.I.; Bandosz, T.J. Role of surface chemistry in adsorption of phenol on activated carbons. *J. Colloid Interf. Sci.* **2003**, *264*, 307–312. [CrossRef]
5. Sun, J.; Liu, X.; Zhang, F.; Zhou, J.; Wu, J.; Alsaedi, A.; Hayat, T.; Li, J. Insight into the mechanism of adsorption of phenol and resorcinol on activated carbons with different oxidation degrees. *Colloids Surf. A Physicochem. Eng. Asp.* **2019**, *563*, 22–30. [CrossRef]
6. Kuśmierek, K.; Białek, A.; Świątkowski, A. Effect of activated carbon surface chemistry on adsorption of phenoxy carboxylic acid herbicides from aqueous solutions. *Desalin. Water Treat.* **2020**, *186*, 450–459. [CrossRef]
7. Kuśmierek, K.; Świątkowski, A.; Skrzypczyńska, K.; Błażewicz, S.; Hryniewicz, J. The effects of the thermal treatment of activated carbon on the phenols adsorption. *Korean J. Chem. Eng.* **2017**, *34*, 1081–1090.
8. Skrzypczyńska, K.; Kuśmierek, K.; Świątkowski, A. Carbon paste electrodes modified with various carbonaceous materials for the determination of 2,4-dichlorophenoxyacetic acid by differential pulse voltammetry. *J. Electroanal. Chem.* **2016**, *766*, 8–15. [CrossRef]
9. Madannejad, S.; Rashidi, A.; Sadeghhassani, S.; Shemirani, F.; Ghasemy, E. Removal of 4-chlorophenol from water using different carbon nanostructures: A comparison study. *J. Mol. Liq.* **2018**, *249*, 877–885. [CrossRef]
10. Atieh, M.A. Removal of phenol from water different types of carbon—a comparative analysis. *APCBEE Procedia* **2014**, *10*, 136–141. [CrossRef]
11. Gogotsi, Y.; Presser, V. *Carbon Nanomaterials*, 2nd ed.; Taylor and Francis Group LLC: Abingdon, UK; CRC Press: Boca Raton, FL, USA, 2014.
12. Béguin, F.; Frąckowiak, E. *Carbons for Electrochemical Energy Storage and Conversion Systems*; Taylor and Francis Group, LLC: Abingdon, UK; CRC Press: Boca Raton, FL, USA, 2010.
13. Biniak, S.; Pakuła, M.; Świątkowski, A.; Bystrzejewski, M.; Błażewicz, S. Influence of high-temperature treatment of granular activated carbon on its structure and electrochemical behavior in aqueous electrolyte solution. *J. Mater. Res.* **2010**, *25*, 1617–1628. [CrossRef]
14. Werengowska-Ciećwierz, K.; Wiśniewski, M.; Terzyk, A.P.; Gurtowska, N.; Olkowska, J.; Kloskowski, T.; Drewa, T.A.; Kiełkowska, U.; Druzyński, S. Nanotube-mediated efficiency of cisplatin anticancer therapy. *Carbon* **2014**, *70*, 46–58. [CrossRef]
15. Childres, I.; Jauregui, L.A.; Park, W.; Cao, H.; Chen, Y.P. Raman spectroscopy of graphene and related materials. In *New Developments in Photon and Materials Research*; Jang, J.I., Ed.; NOVA Science Publishers: New York, NY, USA, 2013; pp. 403–418.
16. Costa, S.; Borowiak-Palen, E.; Kruszyńska, M.; Bachmatiuk, A.; Kaleńczuk, R.J. Characterization of carbon nanotubes by Raman spectroscopy. *Mater. Sci. Pol.* **2008**, *26*, 433–441.
17. Ferrari, A.C.; Robertson, J. Interpretation of Raman spectra of disordered and amorphous carbon. *Phys. Rev. B* **2000**, *61*, 14095–14107. [CrossRef]
18. Sobiesiak, M. Chemical structure of phenols and its consequence for sorption processes. In *Phenolic Compounds—Natural Sources, Importance and Applications*; Soto-Hernández, M., Ed.; IntechOpen: Rijeka, Croatia, 2017; pp. 3–28.
19. Trivedi, N.S.; Kharkar, R.A.; Mandavgane, S.A. 2,4-Dichlorophenoxyacetic acid adsorption on adsorbent prepared from groundnut shell: Effect of preparation conditions on equilibrium adsorption capacity. *Arab. J. Chem.* **2019**, *12*, 4541–4549. [CrossRef]
20. Kuśmierek, K.; Szala, M.; Świątkowski, A. Adsorption of 2,4-dichlorophenol and 2,4-dichlorophenoxyacetic acid from aqueous solution on carbonaceous materials obtained by combustion synthesis. *J. Taiwan Inst. Chem. Eng.* **2016**, *63*, 371–378. [CrossRef]
21. Brito, G.M.; Roldi, L.L.; Schetino, M.Â.; Checon Freitas, J.C.; Cabral Coelho, E.R. High-performance of activated biocarbon based on agricultural biomass waste applied for 2,4-D herbicide removing from water: Adsorption, kinetic and thermodynamic assessments. *J. Environ. Sci. Health B* **2020**, *55*, 767–782. [CrossRef]
22. Kim, T.Y.; Park, S.S.; Kim, S.J.; Cho, S.Y. Separation characteristics of some phenoxy herbicides from aqueous solution. *Adsorption* **2008**, *14*, 611–619. [CrossRef]
23. Białek, A.; Skrzypczyńska, K.; Kuśmierek, K.; Świątkowski, A. Sensitivity change of the modified carbon paste electrodes for detection of chlorinated phenoxyacetic acids. *Chem. Process Eng.* **2019**, *40*, 315–325.
24. Hameed, B.H.; Salman, J.M.; Ahmad, A.L. Adsorption isotherm and kinetic modeling of 2,4-D pesticide on activated carbon derived from date stones. *J. Hazard. Mater.* **2009**, *163*, 121–126. [CrossRef]
25. Njoku, V.O.; Hameed, B.H. Preparation and characterization of activated carbon from corncob by chemical activation with H_3PO_4 for 2,4-dichlorophenoxyacetic acid adsorption. *Chem. Eng. J.* **2011**, *173*, 391–399. [CrossRef]
26. Ramos de Andrade, F.; Alves de Toledo, R.; Manoel Pedro Vaz, C. Electroanalytical methodology for the direct determination of 2,4-dichlorophenoxyacetic acid in soil samples using a graphite-polyurethane electrode. *Inter. J. Electrochem.* **2014**, *2014*, 308926. [CrossRef]
27. Maleki, N.; Safavi, A.; Shahbaazi, H.R. Electrochemical determination of 2,4-D at a mercury electrode. *Anal. Chim. Acta* **2005**, *530*, 69–74. [CrossRef]
28. Prado, A.G.S.; Barcelos, H.T.; Moura, A.O.; Nunes, A.R.; Gil, E.S. Dichlorophenoxyacetic acid anchored on silica-gel modified carbon paste for the determination of pesticide 2,4-D. *Int. J. Electrochem. Sci.* **2012**, *7*, 8929–8939.
29. Xie, C.; Gao, S.; Guo, Q.; Xu, K. Electrochemical sensor for 2,4-dichlorophenoxy acetic acid using molecularly imprinted polypyrrole membrane as recognition element. *Microchim. Acta* **2010**, *169*, 145–152. [CrossRef]
30. Niguso, T.T.; Soreta, T.R.; Woldemariam, E.T. Electrochemical determination of 2,4-dichlorophenoxyacetic acid using bismuth film modified screen-printed carbon electrode. *S. Afr. J. Chem.* **2018**, *71*, 160–165. [CrossRef]

MDPI
St. Alban-Anlage 66
4052 Basel
Switzerland
www.mdpi.com

Materials Editorial Office
E-mail: materials@mdpi.com
www.mdpi.com/journal/materials

Disclaimer/Publisher's Note: The statements, opinions and data contained in all publications are solely those of the individual author(s) and contributor(s) and not of MDPI and/or the editor(s). MDPI and/or the editor(s) disclaim responsibility for any injury to people or property resulting from any ideas, methods, instructions or products referred to in the content.

www.ingramcontent.com/pod-product-compliance
Lightning Source LLC
LaVergne TN
LVHW070045120526
838202LV00101B/633